M336
Mathematics and Computing: a third-level course

GROUPS & GEOMETRY

UNIT GE1
COUNTING WITH GROUPS

Prepared for the course team by
Roy Nelson

The Open University

This text forms part of an Open University third-level course.
The main printed materials for this course are as follows.

Block 1
Unit IB1	Tilings	
Unit IB2	Groups: properties and examples	
Unit IB3	Frieze patterns	
Unit IB4	Groups: axioms and their consequences	

Block 2
Unit GR1	Properties of the integers	
Unit GR2	Abelian and cyclic groups	
Unit GE1	Counting with groups	
Unit GE2	Periodic and transitive tilings	

Block 3
Unit GR3	Decomposition of Abelian groups	
Unit GR4	Finite groups 1	
Unit GE3	Two-dimensional lattices	
Unit GE4	Wallpaper patterns	

Block 4
Unit GR5	Sylow's theorems	
Unit GR6	Finite groups 2	
Unit GE5	Groups and solids in three dimensions	
Unit GE6	Three-dimensional lattices and polyhedra	

The course was produced by the following team:

Andrew Adamyk (BBC Producer)
David Asche (Author, Software and Video)
Jenny Chalmers (Publishing Editor)
Bob Coates (Author)
Sarah Crompton (Graphic Designer)
David Crowe (Author and Video)
Margaret Crowe (Course Manager)
Alison George (Graphic Artist)
Derek Goldrei (Groups Exercises and Assessment)
Fred Holroyd (Chair, Author, Video and Academic Editor)
Jack Koumi (BBC Producer)
Tim Lister (Geometry Exercises and Assessment)
Roger Lowry (Publishing Editor)
Bob Margolis (Author)
Roy Nelson (Author and Video)
Joe Rooney (Author and Video)
Peter Strain-Clark (Author and Video)
Pip Surgey (BBC Producer)

With valuable assistance from:

Maths Faculty Course Materials Production Unit
Christine Bestavachvili (Video Presenter)
Ian Brodie (Reader)
Andrew Brown (Reader)
Judith Daniels (Video Presenter)
Kathleen Gilmartin (Video Presenter)
Liz Scott (Reader)
Heidi Wilson (Reader)
Robin Wilson (Reader)

The external assessor was:
Norman Biggs (Professor of Mathematics, LSE)

The Open University, Walton Hall, Milton Keynes, MK7 6AA.

First published 1994. Reprinted 1997, 2002, 2005, 2009.

Copyright © 1994 The Open University

All rights reserved. No part of this publication may be reproduced, stored in a retrieval system or transmitted in any form or by any means, without written permission from the publisher or a licence from the Copyright Licensing Agency Limited. Details of such licences (for reprographic reproduction) may be obtained from the Copyright Licensing Agency Ltd of 90 Tottenham Court Road, London, W1P 9HE.

Edited, designed and typeset by the Open University using the Open University TeX System.

Printed in Malta by Gutenberg Press Limited.

ISBN 0 7492 2169 0

This text forms part of an Open University Third Level Course. If you would like a copy of *Studying with the Open University*, please write to the Central Enquiry Service, PO Box 200, The Open University, Walton Hall, Milton Keynes, MK7 6YZ. If you have not already enrolled on the Course and would like to buy this or other Open University material, please write to Open University Educational Enterprises Ltd, 12 Cofferidge Close, Stony Stratford, Milton Keynes, MK11 1BY, United Kingdom.

1.3

The paper used for this book is FSC-certified and totally chlorine-free. FSC (the Forest Stewardship Council) is an international network to promote responsible management of the world's forests.

CONTENTS

Study guide		4
Introduction		5
1	**Equivalent colourings**	**6**
	1.1 Equivalence classes and partitions	6
	1.2 Orbits	8
2	**Group actions**	**9**
	2.1 The symmetry group of the square picture frame	9
	2.2 Group actions on colourings	10
3	**The Counting Lemma**	**13**
	3.1 Counting orbits	13
	3.2 Stabilizers	15
	3.3 Inert pairs	17
	3.4 Exercises on the Counting Lemma	19
	3.5 The Orbit–stabilizer Theorem	20
4	**The cycle index**	**22**
	4.1 Permutations and cycles	22
	4.2 The cycle symbol	23
	4.3 Counting colourings	25
5	**Pólya's enumeration formula**	**26**
	5.1 Listing the colourings	26
	5.2 Pólya's Theorem	29
Appendix: colouring the cube		33
Solutions to the exercises		35
Objectives		46
Index		47

STUDY GUIDE

This is the first unit in the Geometry stream. It should be studied before the other geometry unit in this block, *Unit GE2*, but it is up to you whether you study the Geometry stream units before or after the Groups stream units in the block. (Another possibility would be to study them alternately.)

Sections 1 and 2 develop the concept of group actions (which you met in *Unit IB2*), in the context of examining the possible ways of colouring a geometric object. They should prove to be straightforward.

Sections 3, 4 and 5 go on to use group actions to develop a powerful method of counting and listing the 'essentially different' colourings of such an object, subject to various conditions. These sections are rather more demanding than Sections 1 and 2.

The video programme associated with this unit is VC2A, the first programme on the second video-cassette. It is, in a sense, an overview of the unit — an application of the whole unit to the problem of the colourings of a hexagonal pyramid. It could be viewed at any time during your study of the unit, but we suggest that you might like to treat it as an introduction.

There is no audio programme associated with this unit.

You will not require the *Geometry Envelope* in your study of this unit.

INTRODUCTION

When faced with a collection of objects, we can usually count them by pointing to the objects one at a time and 'calling them off' one, two, three, four, ..., and so on.

Sometimes, if the number of objects is large, we may look for some convenient short cut: for instance, you might count the number of stars in the famous Stars and Stripes flag by some method such as observing that there are five rows of six stars (giving 30 stars) and four rows of five stars (giving a further 20 stars) and hence 50 stars in all (see Figure 0.1).

Figure 0.1

Even though you used a short cut, you can probably still *see* what you are counting: you are still able to count the objects in this simple way if need be. But what do you do if you cannot tell which objects are to be counted? You cannot *point* to the objects one at a time if you are not sure which objects to count and which to ignore.

Suppose you wanted to count the number of ways of colouring the faces of a cube with three colours, each face being red or white or blue. With three choices for each of the six faces, there would seem to be $3^6 = 729$ possible colourings; but should you really count all of these? One of them might have, say, blue on the top face and white on all the other faces; another might have blue on the front face and white everywhere else; others will have blue on some other face and white elsewhere. Surely you should count only one of these, because you will not be able to tell them apart once they are moved about?

The problem is to do with the *symmetries* of the cube — and the solution (as you might expect!) is to be found in an appeal to groups. This particular problem is rather more complicated than the others in the unit: you will find it solved in the Appendix.

In Section 1, we study the way in which the symmetries of, say, a picture frame lead us to regard two colourings as the same.

In Section 2, we expand the concept of a *group action*, which you met in *Unit IB2*.

In Section 3, we make use of the group of symmetries to help us to count the number of distinct colourings.

In Section 4, by examining the cycle structure of the elements of the symmetry group, we derive a powerful technique for counting such elusive 'objects' as distinct colourings.

In Section 5, a further development of the technique enables us both to count and actually to list the distinct colourings. At the end of this section *Pólya's Theorem* helps us to solve the problem of classifying colourings of geometric configurations.

1 EQUIVALENT COLOURINGS

If you have a collection of objects and want to know how many different objects there are, you face the often difficult problem of deciding when two objects are to be regarded as different: what precisely do we mean by saying that two objects are *the same*?

1.1 Equivalence classes and partitions

The Introduction described the problem of counting the number of different ways of colouring the faces of a cube when each face is red or white or blue. Let's look at a simpler problem of the same type.

Example 1.1

Figure 1.1 shows a number of ways of colouring a square picture frame, where each side of the frame is either black or white.

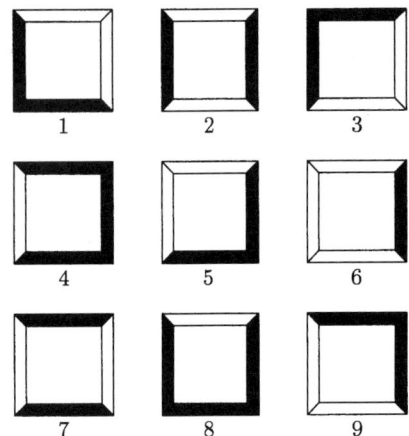

Figure 1.1

Which of the picture frames in Figure 1.1 do you think are 'the same', in some sense?

Clearly there is a sense in which they are all different. If you made nine copies of your passport photograph and stuck one, the usual way up, into each of the frames, and then cut out all the framed photographs, you would certainly be able to tell them apart. On the other hand, suppose you were to cut out the frames *without* sticking pictures into them, shake them up in a bag, and then pick one out. If the selected frame had two adjacent black sides and two adjacent white sides, then you would not be able to tell whether it was number 1, 3, 5 or 9. These four frames are 'the same' in the sense that rotating any one of them by quarter-turns will produce the others. You don't know how much rotation they have undergone in the bag, which is why you can't tell which one of the four is the one you selected. ♦

Exercise 1.1

Counting two frames as 'the same' if one can be converted into the other by rotation, how many 'different' picture frames are there in Figure 1.1?

What we have effectively done is to partition frames 1 to 9 into *equivalence classes*; two frames are in the same class if one can be rotated to produce the other or, equivalently, if once they have been cut out and shuffled in a bag, they cannot be distinguished from each other. The objects which we are partitioning into classes are frames in which we know the colours of the top, bottom, left and right sides *individually*. The equivalence classes are classes within which we know only the *relative* positions of the colours.

Example 1.2

It is now worth extending Example 1.1 and counting *all* the classes of equivalent colourings of our square picture frame, using black or white sides. The top, bottom, left and right sides can each be individually coloured black or white, so the number of 'objects', or individual colourings, is $2^4 = 16$. We are allowing rotations about the centre of the frame by integer multiples of $\pi/2$, and counting two objects as being in the same class if one can be transformed into the other by means of such an operation. Therefore, there are six equivalence classes, as shown in Figure 1.2.

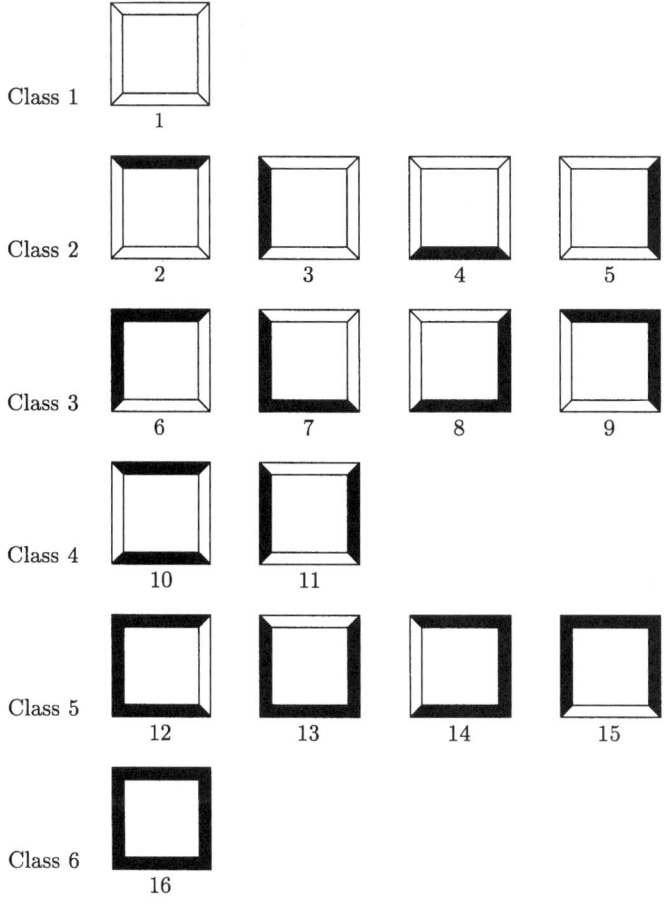

Figure 1.2

♦

Now test your understanding of what we have done so far by trying the following exercise.

Exercise 1.2

Consider a hexagon H divided into three rhombuses, each of which is to be coloured with a single colour (see Figure 1.3). Using the colours black and white, and allowing rotations through any multiple of $2\pi/3$, find the individual colourings and the classes of equivalent colourings.

Figure 1.3

A rhombus is a four-sided figure with sides of equal length (and whose diagonals therefore intersect at right angles).

Note that Figure 1.3 is a picture of a flat hexagon, *not* a cube. There are several pictures of hexagons in this unit, but no pictures of cubes except in the Appendix.

1.2 Orbits

So far in this unit, the 'allowable transformations' have all been rotations about the centre of the figures. In the examples we have considered, it would not have mattered if we had 'allowed' reflections as well. But, in other circumstances, allowing transformations to include reflections may well make a difference. For example, if the colours black, grey and white are used to colour the picture frames, are the colourings in Figure 1.4 equivalent?

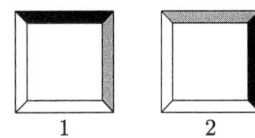

Figure 1.4

One colouring cannot be transformed into the other by rotation in the plane of the paper — but if the frames are made of flat pieces of wood, painted the same colour on both sides, then turning number 1 over will produce number 2. (In terms of isometries of the plane, the effect is of reflection in the diagonal passing through the top right and bottom left corners of the frame.)

Thus, we cannot tell how many equivalence classes of picture frames we can produce using three colours, until we know whether or not we are allowed to turn the frames over. If we denote by F a plane drawing of an unpainted frame (Figure 1.5), then the question becomes: are the 'allowable' transformations the direct symmetry group $\Gamma^+(F)$, or are they the full symmetry group $\Gamma(F)$? Neither of these specifications is mathematically more 'right' than the other. If the frames cannot be turned over, we use the group $\Gamma^+(F)$ in the mathematical specification, while if they can, we use $\Gamma(F)$.

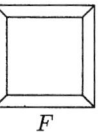

Figure 1.5

$\Gamma(F)$ and $\Gamma^+(F)$ are defined in *Unit IB3*.

Notice that the effect of any of the allowable transformations upon a particular colouring is to produce another colouring in the same equivalence class. Indeed, if we take any colouring and apply to it the allowable transformations, we obtain all the colourings in the same equivalence class. Thus (for instance) starting with the colouring labelled 3 in Figure 1.2, the four allowable rotations $e, r[\pi/2], r[\pi], r[3\pi/2]$ produce respectively the colourings 3, 4, 5, 2, which constitute the whole of Class 2.

Exercise 1.3

Consider again the hexagon H of Exercise 1.2, and suppose that each rhombus is to be coloured black or grey or white. One of the possible colourings is given in Figure 1.6.

(a) Suppose the group of allowable transformations is
$\Gamma^+(H) = \{e, r[2\pi/3], r[4\pi/3]\}$. Apply these transformations, and thus find the equivalence class of the colouring in Figure 1.6.

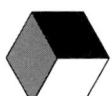

Figure 1.6

(b) Repeat the process on the assumption that the group of allowable transformations is

$$\Gamma(H) = \{e, r[2\pi/3], r[4\pi/3], q[0], q[\pi/3], q[2\pi/3]\}.$$

In effect, what we are doing is to allow the group of allowable transformations to *act* on the set of colourings, and to take the *orbits* under this action to be the equivalence classes which we wish to count. You met group actions and orbits in Section 5 of *Unit IB2*, and in the next section of this unit we shall revise these concepts and apply them in detail to colourings of geometric objects.

2 GROUP ACTIONS

2.1 The symmetry group of the square picture frame

Let F be the square picture frame of Figure 1.5. We may describe its symmetry group, $\Gamma(F)$, in a way analogous to that used in the case of the regular hexagon, which was considered in Section 1 of *Unit IB2*. The group is referred to as D_4, and is generated by $r = r[\pi/2]$ and $s = q[0]$:

You met D_4 in *Unit IB4*.

$$D_4 = \{r^m s^n : m = 0, \ldots, 3, \ n = 0, 1; \ r^4 = s^2 = e, \ sr = r^3 s\}.$$

We may record the effect of each of the elements on F by a device which you may well have seen in your previous mathematical studies. We draw a small identification symbol in each of the four sides; then, when a symmetry is applied, the positions of these symbols allow us to keep track of which side has been mapped where. Figure 2.1 shows the effect of the symmetry r.

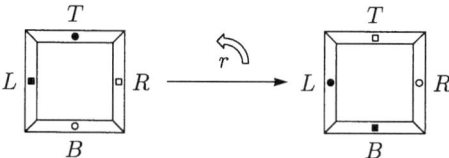

Figure 2.1

Note that the *positions* T (top), L (left), B (bottom) and R (right) remain fixed, whereas the *identification symbols* show us that, in this case, the side that *was* in position T has been mapped to position L and that which *was* in position L has been mapped to B, and so on.

Doing the same for each of the eight symmetries, we can construct a table describing the effect of the symmetries on the sides, as shown in Table 2.1.

	T	L	B	R
e	T	L	B	R
r	L	B	R	T
r^2	B	R	T	L
r^3	R	T	L	B
s	B	L	T	R
rs	R	B	L	T
r^2s	T	R	B	L
r^3s	L	T	R	B

Table 2.1

At the risk of being tedious, we repeat: the *positions* T, L, B, R do not move. The fact that, for example, the table entry in row rs and column L is B means that the symmetry rs maps the side that *was* in position L, into position B. Mathematically we may consider Table 2.1 as defining a *group action*.

We remind you here of the characterization of group action that was given in Section 5 of *Unit IB2*, immediately after the formal definition.

> ***Definition 2.1 Group action***
>
> A group G is said to **act** on a set X if, for each $g \in G$ and $x \in X$:
> (a) $g \wedge x \in X$, for all $g \in G$, $x \in X$;
> (b) $e \wedge x = x$, for all $x \in X$, where e is the identity element of G;
> (c) $(gh) \wedge x = g \wedge (h \wedge x)$, for $g, h \in G$ and $x \in X$.

This definition is not the same as Definition 5.1 in *Unit IB2*. It corresponds to Lemma 5.1, which uses the $\wedge$ notation and which provides an alternative, and equivalent, definition of a group action.

In this case, $G = D_4$ and X is the set of positions $\{T, L, B, R\}$. We call such a set of positions a **configuration**. We *define* the action by saying that $g \wedge x$ is the position to which the side of F that starts in position x is moved by the symmetry g. The corresponding table (Table 2.1 in this case) is called a **group action table**.

Note the difference between a group action table and a Cayley table! Table 2.2 is the Cayley table for D_4.

	e	r	r^2	r^3	s	rs	r^2s	r^3s
e	e	r	r^2	r^3	s	rs	r^2s	r^3s
r	r	r^2	r^3	e	rs	r^2s	r^3s	s
r^2	r^2	r^3	e	r	r^2s	r^3s	s	rs
r^3	r^3	e	r	r^2	r^3s	s	rs	r^2s
s	s	r^3s	r^2s	rs	e	r^3	r^2	r
rs	rs	s	r^3s	r^2s	r	e	r^3	r^2
r^2s	r^2s	rs	s	r^3s	r^2	r	e	r^3
r^3s	r^3s	r^2s	rs	s	r^3	r^2	r	e

Table 2.2

The Cayley table is made up entirely of elements of the group itself, and is always square, whereas the elements in the body of the group action table are (in this case) the positions T, L, B, R. This group action table (i.e. Table 2.1) is not square, as there are eight group elements but only four positions in the configuration.

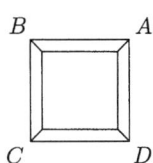

Figure 2.2

Exercise 2.1

Label the positions of the corners of F, as in Figure 2.2, and draw up a group action table for the action of D_4 on $\{A, B, C, D\}$.

Exercise 2.2

Draw up a group action table for the action of D_4 on the configuration consisting of the diagonals, AC and BD, of F (Figure 2.3).

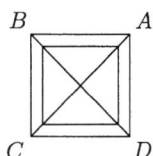

Figure 2.3

So much for group actions on configurations. What we are really interested in, though, is how the group acts on *colourings*: this is the subject of the next subsection.

2.2 Group actions on colourings

One of the difficulties we shall encounter in dealing with group actions on colourings is that there are three sets to consider:

- the group itself, say G;
- the configuration on which the group acts, say X;
- the set of colourings of X, say C.

As well as the three sets, there are two group actions to think about: we assume that G acts on X, but we shall usually be concerned with the action of G on the set C of colourings. This arises naturally from the action of G on X: each element of G acts on the elements of X in a particular way and, in doing so, it affects any colouring of the configuration X to produce another colouring. In general, this will be a different colouring but it may sometimes be the same. Let us look again at the simple example in Exercise 1.2.

Example 2.1

We consider a hexagon divided into three rhombuses (see Figure 2.4). We have labelled the positions of the rhombuses: 1, 2, 3.

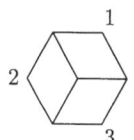

Figure 2.4

The transformations allowed are rotations through any multiple of $2\pi/3$, simply because such rotations are the only ones that permute the rhombuses among themselves.

Here the configuration X is the set $\{1, 2, 3\}$ of rhombus positions. The group G acting on X is

$$G = \{e, r, r^2 : r^3 = e\},$$

where r is the rotation through $2\pi/3$ about the centre of the hexagon.

On this occasion, the group action table (Table 2.3) and the Cayley table (Table 2.4) are both square:

	1	2	3
e	1	2	3
r	2	3	1
r^2	3	1	2

	e	r	r^2
e	e	r	r^2
r	r	r^2	e
r^2	r^2	e	r

Table 2.3 Group action table   Table 2.4 Cayley table

What about the *colourings* of the rhombuses, using black and white? In Exercise 1.2 we found that there are eight of these; calling this set C, we may label them individually c_1 to c_8. They are shown, collected into equivalence classes, in Figure 2.5.

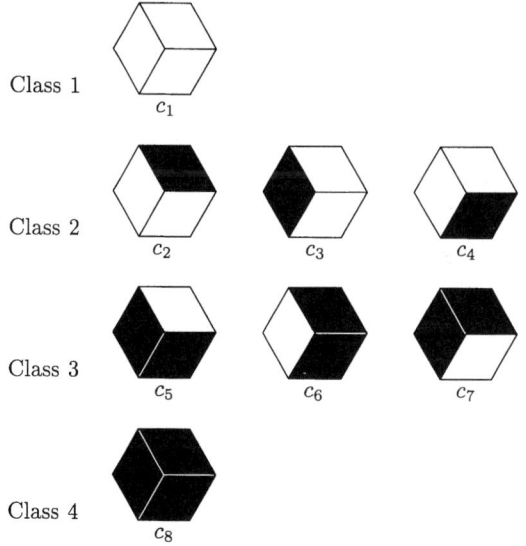

Figure 2.5

The group elements permute the colourings among themselves within each class. Thus, for example, the element r^2 moves c_5 to c_7. ◆

Before you attempt the following exercises, we remind you of the formal definition of an orbit, from *Unit IB2*.

> **Definition 2.2 Orbit**
>
> If a group G acts on a set X and x is an element of X, then the **orbit** of x under G is the set of elements of X obtained by acting on x with the elements of G. It is denoted by
>
> $\text{Orb}(x) = \{g \wedge x : g \in G\}.$

Exercise 2.3

Construct the group action table for the action of the group G on the set of colourings, $C = \{c_1, c_2, \ldots, c_8\}$, and list the orbits.

This exercise illustrates the important point that the equivalence classes of colourings are just the orbits of the group action on the individual colourings. There are, as we have seen, just four essentially different ways, given two colours, of colouring the three rhombuses into which the hexagon is divided; two colourings are equivalent precisely when they are in the same orbit under the action of the group of direct symmetries of the hexagon.

Now it is time for you to try your hand at all this.

In these two exercises the set X consists of the four quadrants of a square S. The positions of the quadrants are labelled A, B, C, D, as shown in Figure 2.6. The group G is D_4, and C is the set of all the colourings of X in which exactly two quadrants are black and two are white.

Note that this is *not* the set of *all sixteen* quadrant colourings using black and white.

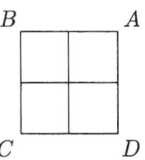

Figure 2.6

Exercise 2.4

Identify the group action table for the group D_4 acting on X.

Exercise 2.5

Sketch the six colourings comprising C. Label them $c_1, \ldots, c_6$, then write out the group action table for the group D_4 acting on C. Finally, list the orbits and describe them in your own words.

You may say that you could easily have arrived at the conclusion that there are only these two essentially different ways of colouring two quadrants black and two white without all that labour! Of course you could; but could you do the same if you had to colour, say, the faces of a dodecahedron using, perhaps, five colours? The next section shows you how, using group actions, you could work out the answer in even the most complicated cases.

3 THE COUNTING LEMMA

Some of the problems we have been looking at may have been easy, some perhaps harder, but that we were able to do them at all is due solely to their being relatively small and not too complex. It would be entirely another matter if, for instance, we were to tackle the problem of colouring the faces of a dodecahedron using five colours: we should then be faced with the task of deciding which of nearly a quarter of a billion colourings are equivalent. We need to employ some more sophisticated technique than brute force, and for this we turn to group theory — to group actions. Of course, we have already spoken of group actions, but we have not yet seen how they can be used to assist in the task of counting equivalence classes.

3.1 Counting orbits

We shall start with our familiar picture frame, but this time we shall use three colours.

Example 3.1

The configuration X is again the set $\{T, L, B, R\}$ of the four positions of the edges of a square picture frame. Let us assume that the frame cannot be turned over, so the symmetry group is $G = \{e, r, r^2, r^3\}$ where $r = r[\pi/2]$. This time we are given three colours: black, grey and white. We ask: what are the classes of equivalent colourings *if all three colours have to be used*? That is, what are the orbits under the action of G on this set of colourings?

This problem means that *one* colour must appear on *two* of the sides and then the other two sides can be coloured two ways: this gives twelve objects for each choice of 'two-sided' colour. Figure 3.1 shows this when two of the sides are black.

From now on, the individual colourings on which a symmetry group acts will often be referred to as *objects*.

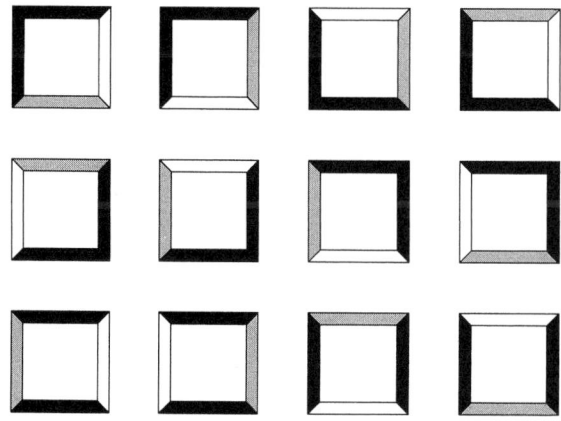

Figure 3.1

Similarly, by considering first grey and then white as the colours of two of the sides, we obtain 12 objects each, thus there are 36 objects in total.

Now, the orbits are *all* of size 4, since *every* non-trivial rotation sends *any* of the above colourings to a different one. Thus the number of orbits is $36/4 = 9$. ♦

Example 3.2

The configuration, the symmetry group and the colour set remain as for Example 3.1, but this time we ask: what are the orbits *if exactly two of the three colours must be used*?

There are fourteen objects using the colour pair {black, grey}, fourteen using {black, white} and fourteen using {grey, white}, giving a total of 42 objects.

This time, however, not all the orbits are the same size. Considering just the objects using the colour pair {black, grey}, there are twelve objects each of which is mapped to a different one by *any* non-trivial rotation, giving the three equivalence classes (orbits) shown in Figure 3.2.

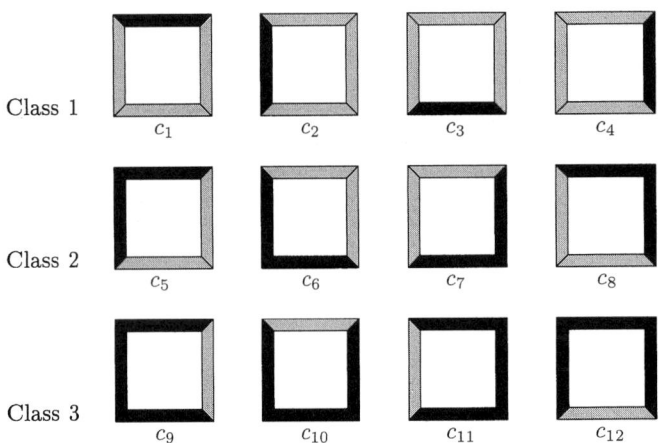

Figure 3.2

The other two black and grey objects behave differently; they map to each other under r or r^3, but *each maps to itself* under r^2, as Figure 3.3 shows.

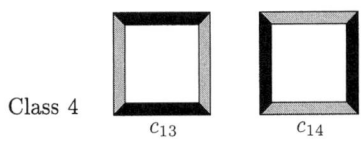

Figure 3.3

Thus, considering just the black and grey objects, there are $12/4 = 3$ orbits of the first type and $2/2 = 1$ of the second type. When we add in the black and white objects and the grey and white objects, we find that there are $36/4 = 9$ orbits of the first type and $6/2 = 3$ of the second type, making 12 orbits in all. Thirty-six of the objects belong to orbits that have no non-trivial rotational symmetry, and we are required to divide by 4 to count these orbits, while the others belong to orbits that have rotational symmetry of order 2, and we are required to divide by 2 to count these.

Another way to look at this would be to count each of the latter objects *twice*, as we find them, to allow for their symmetries: that is, we would have $36 + (2 \times 6) = 48$ 'effective objects' — which divides by 4 to get the correct number of 12 orbits. ♦

Exercise 3.1

How many orbits (equivalence classes) are there if *at least two* of the three colours must be used?

Example 3.3

The configuration, the symmetry group and the colour set remain as for Examples 3.1 and 3.2, but now we ask: what are the orbits if *just one colour is used*?

There are three objects, as shown in Figure 3.4.

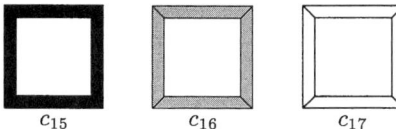

c_{15} c_{16} c_{17}

Figure 3.4

Each of these remains unchanged under *all four* elements of the symmetry group. Thus, in the same way that we counted *twice* each of the objects having rotational symmetry of order 2 so we should count each of the uniformly coloured objects *four* times. So we have $3 \times 4 = 12$ 'effective objects', which we divide by 4 again to obtain the correct number of 3 orbits. ♦

Example 3.4

The configuration, the symmetry group and colour set remain as before, and we now ask: how many orbits are there altogether?

We obtain the answer by adding together the answers that we obtained in Examples 3.1–3.3. Thus, Example 3.1 tells us that there are 36 'effective objects' using all three colours; Example 3.2 shows that there are 48 'effective objects' using exactly two colours; and Example 3.3 gives 12 'effective objects' using just one colour. Thus the total number of 'effective objects' is $36 + 48 + 12 = 96$, giving $96/4 = 24$ orbits. ♦

3.2 Stabilizers

In the examples of Subsection 3.1, we counted a particular object (i.e. a particular colouring of the picture frame) once if it had trivial rotational symmetry (i.e. rotational symmetry of order 1), twice if it had rotational symmetry of order 2, and four times if it had rotational symmetry of order 4. That is, we 'weighted' each object in accordance with its symmetries. How can we describe a *general* counting method based on this weighting idea? Well, we multiplied each object *by the number of elements of G under whose action it did not change*, i.e. we weighted each object by the size of its stabilizer. Then we counted orbits (equivalence classes) by adding these weights and dividing by the order of G.

You have already seen the definition of a **stabilizer**, but we repeat it here for easy reference.

See *Unit IB2*.

Definition 3.1 Stabilizer

If a group G acts on any set X, and x is an element of X, then the **stabilizer** of x under G is the set of elements of G which fix x. It is denoted by

$$\mathrm{Stab}(x) = \{g : g \in G,\ g \wedge x = x.\}$$

This is a general definition, valid for any group action, though in this unit X will always be a set C of colourings of some configuration.

The method of counting, which we introduced above, weights each object by the size of its stabilizer, and then gives:

$$\text{number of orbits} = \frac{1}{|G|} \sum_{c \in C} |\text{Stab}(c)|. \qquad (3.1)$$

> We have not yet *proved* that this counting method gives the right answer — merely noted that it appears reasonable. The proof is given in Subsection 3.5.

This is fine, as far as it goes, but we still seem to need a good knowledge of each of the orbits we propose to count, before we can actually count them. So does this really help us to count, say, the colourings of the dodecahedron, when five colours are available?

Not yet, but we do have another trick up our sleeve — one that even a member of the Magic Circle would be proud of! It has been associated with the names of at least three great mathematicians: Cauchy, Frobenius and Burnside. Many mathematicians call it Burnside's Lemma, and many others call it the Cauchy–Frobenius Lemma. We shall preserve neutrality and call it the *Counting Lemma*.

When we say that a colouring such as c_1 in Figure 3.2 has no rotational symmetry, what we really mean is that, so far as the action by the group G of rotations is concerned, the stabilizer is just $\{e\}$:

$$\text{Stab}(c_1) = \{e\}.$$

So let us count it by the symbol

$$(c_1, e).$$

On the other hand, c_{13} in Figure 3.3 has as its stabilizer the set $\{e, r^2\}$:

$$\text{Stab}(c_{13}) = \{e, r^2\}.$$

So let us count it by the *two* symbols

$$(c_{13}, e), (c_{13}, r^2);$$

and when we count c_{15} of Figure 3.4, we shall of course use the *four* symbols

$$(c_{15}, e), (c_{15}, r), (c_{15}, r^2), (c_{15}, r^3).$$

So, the total number of orbits (equivalence classes) is the *number of all these symbols*, divided by $|G|$.

Now for the Magic Circle contribution! These symbols can be *arranged differently*. We can count all those whose group element is e, then all those whose group element is r, then r^2 ... and so on. In this way, each group element is counted once for *each* element of X which it leaves *fixed*, so it is the *group elements* which are now to be weighted — and the summation is carried out *over the group*, which we know more about.

Example 3.5

Let us now check this for Example 1.2, in which the colouring of four sides of a picture frame either black or white was investigated.

For each group element in turn (the four rotations), we list the objects (i.e. the colourings) which it does not alter.

The identity, e, does not alter any colouring; so against it we list all sixteen of the colourings in Figure 1.2.

$$e:$$

$$(c_1, e), (c_2, e), (c_3, e), (c_4, e), (c_5, e), (c_6, e), (c_7, e), (c_8, e),$$
$$(c_9, e), (c_{10}, e), (c_{11}, e), (c_{12}, e), (c_{13}, e), (c_{14}, e), (c_{15}, e), (c_{16}, e).$$

> The colouring captioned i in Figure 1.2 is denoted here by c_i ($i = 1, 2, \ldots, 16$).

The rotation r alters every colouring except the all black and all white colourings; these are the only two it does *not* alter (see Figure 3.5).

r :

$(c_1, r), (c_{16}, r).$

Note that this is a different example from that of Figures 3.2, 3.3 and 3.4, so c_1 and c_{16} are *not* the same objects as in those figures.

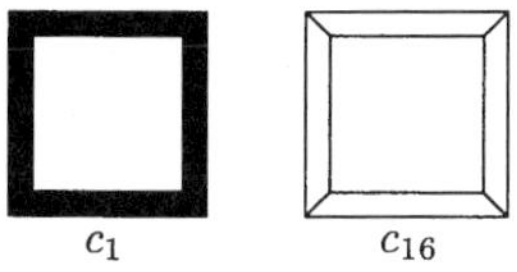

Figure 3.5

There are four colourings unaltered by r^2 (see Figure 3.6).

r^2 :

$(c_1, r^2), (c_{10}, r^2), (c_{11}, r^2), (c_{16}, r^2).$

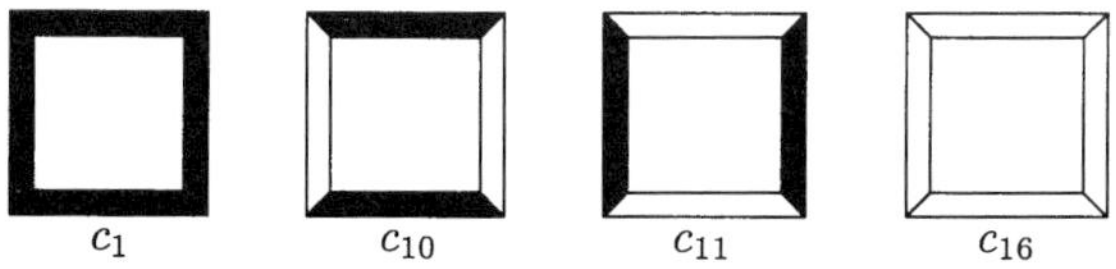

Figure 3.6

Finally, the rotation r^3 behaves just like r, and the only two colourings that it does not alter are all black and all white.

r^3 :

$(c_1, r^3), (c_{16}, r^3).$

Altogether, the sum of the sizes of the sets of objects above is $16 + 2 + 4 + 2 = 24$. Dividing by the order of the group, we get $24/4 = 6$, which is precisely the number of equivalence classes (orbits) which we found in Section 1. ♦

3.3 Inert pairs

We shall now look at this counting process another way.

As we have seen, the number of orbits can be calculated by counting all the pairs (c_i, g_j) for which the colouring c_i is fixed by the group element g_j, and then dividing the total number of such pairs by $|G|$, the order of the group. The following definition will help.

Definition 3.2 Inert pairs

When a group G acts on a set X, then the **inert pairs** of the action are the ordered pairs (x, g) such that $g \wedge x = x$.

Once again, this definition is valid for *any* group action, but in this unit X will always be a set C of colourings of some configuration.

We can then express our result above as:

$$\text{number of orbits} = \frac{1}{|G|} |\text{set of inert pairs}|. \qquad (3.2)$$

Of course, it's one thing to talk about the total number of inert pairs, but it's quite another to do the counting! These inert pairs are surely even more elusive than the colourings ... for they are a combination of a colouring *and* a group element!

But let's persevere with the idea and see where it takes us. It sometimes helps to draw a picture (see Figure 3.7).

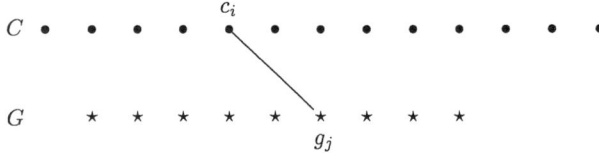

Figure 3.7

At the top, we show a set of points representing the individual colourings. At the bottom, another set of points is used to represent the elements of the group G.

Lines drawn between the two sets of points represent the inert pairs (c, g): so the line from c_i to g_j indicates that the group element g_j leaves the colouring c_i fixed, and each line corresponds to an inert pair.

Now, it must at first sight seem more difficult to count the number of lines than to count the number of orbits directly. Yet there is a magic trick to all this — the Counting Lemma. It arises from the disarmingly simple observation that the number of lines can be found by counting the number of ends, *either* at the top (which we have decided may be difficult) *or* at the bottom — which may be easier, since we are then in the group G, which we *do* know something about.

Now, consider the number of 'ends' at one particular group element g — that is, the number of inert pairs (c_i, g).

Definition 3.3 Fixed set

Let a group G act on a set C. Then for each element g of G, the **fixed set** of g is

$$\text{Fix}(g) = \{c \text{ in } C \text{ such that } g \text{ leaves } c \text{ fixed}\}.$$

If we were able to count the *top* ends of the lines, we would get

$$\sum_{c \in C} |\text{Stab}(c)|.$$

Counting the *bottom* ends, we get

$$\sum_{g \in G} |\text{Fix}(g)|.$$

Thus, of course, as each of these is the number of lines, or inert pairs, we have

$$\sum_{c \in C} |\text{Stab}(c)| = \sum_{g \in G} |\text{Fix}(g)|.$$

Dividing through by $|G|$, we obtain

$$\frac{1}{|G|} \sum_{c \in C} |\text{Stab}(c)| = \frac{1}{|G|} \sum_{g \in G} |\text{Fix}(g)|. \tag{3.3}$$

Now we saw in Equation 3.1 that the expression on the left of Equation 3.3 gives the number of orbits, or equivalence classes, of colourings — which is precisely what we want. Hence the expression on the right also gives the number of orbits. This result is the Counting Lemma. Of course, the above argument does not depend on the set being a set of colourings; it works for *any* action by a finite group G on *any* set X.

> **Theorem 3.1 Counting Lemma**
>
> Let G be any finite group acting on any set X. Then
> $$\text{Number of orbits} = \frac{1}{|G|} \sum_{g \in G} |\text{Fix}(g)|.$$

The proof of this result requires, first of all, that we prove the validity of the rather vague arguments that led to Equation 3.1 in Subsection 3.2. We shall defer this proof to the end of the section; a more important priority is for you to gain some practice in the use of the Counting Lemma in counting equivalence classes of colourings of various configurations. The exercises in the next subsection are designed with this in mind.

3.4 Exercises on the Counting Lemma

Exercise 3.2

(a) Apply the Counting Lemma to count the number of equivalence classes of colourings of a square picture frame each of whose sides must be coloured black or white, assuming that the frame cannot be turned over.

(b) Repeat the count under the assumptions that the sides are the same colour on both surfaces and that the frame *can* be turned over.

Exercise 3.3

Repeat Exercise 3.2 (both parts) for a square divided into quadrants, each of which is to be coloured black or white.

Exercise 3.4

Repeat Exercise 3.2 for a rectangular picture frame (see Figure 3.8).

Exercise 3.5

Repeat Exercise 3.3 for a rectangle divided into quadrants (see Figure 3.9).

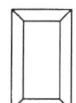

Figure 3.8   *Figure 3.9*

Exercise 3.6

A regular hexagon is divided into three rhombuses (see Figure 3.10). Assuming that the hexagon cannot be turned over, how many equivalence classes of colourings are there in which each rhombus is coloured black, grey or white?

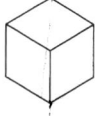

Figure 3.10

Exercise 3.7

Repeat Exercise 3.6, assuming that the hexagon is made of a transparent material and can be turned over.

Exercise 3.8

Suppose you were to draw four $1\,\text{cm} \times 1\,\text{cm}$ squares, and colour each of them as in Figure 3.11.

(a) Assuming that you use ordinary paper and colour only one side, how many equivalence classes are there of ways of assembling your four $1\,\text{cm} \times 1\,\text{cm}$ squares into a $2\,\text{cm} \times 2\,\text{cm}$ square?

(b) Repeat the calculation, assuming that a transparent material has been used.

Figure 3.11

Exercise 3.9

This is a little harder! Suppose we have a square picture frame that can be turned over, and we paint each side either black or white, but the upper and lower surfaces of each side are painted *independently*, so that they can either be the same colour or opposite colours. (Thus the arrangement of colours which we see when the frame is one way up, is no guide at all to what we shall see when we turn it over.) How many equivalence classes of colourings are there now?

3.5 The Orbit–stabilizer Theorem

In Subsection 3.2, we described a counting method which (we noted) seemed reasonable but which we had not actually proved. We repeat it here:

$$\text{number of orbits} = \frac{1}{|G|} \sum_{c \in C} |\operatorname{Stab}(c)|. \tag{3.1}$$

We shall now show that this is a direct consequence of a result which you may have come across in your previous mathematical studies, namely the *Orbit–stabilizer Theorem*.

Theorem 3.2 Orbit–stabilizer Theorem

When a finite group G acts on a finite set X, then for each $x \in X$:

$$|\operatorname{Orb}(x)| \times |\operatorname{Stab}(x)| = |G|.$$

Proof

Suppose

$$\operatorname{Orb}(x) = \{x_1, x_2, \ldots, x_m\} \quad \text{(with } x = x_1\text{)}.$$

For each x_i, select one particular element g_i of G such that

$$g_i \wedge x = x_i.$$

(There must be at least one available for each i, as $x_i \in \operatorname{Orb}(x)$.)

Suppose

$$\operatorname{Stab}(x) = \{h_1, h_2, \ldots, h_n\}.$$

Now let g be *any* element of G. Then $g \wedge x$ is exactly one of the elements of $\operatorname{Orb}(x)$, say x_k.

Form the group element $g_k^{-1} g$, and consider how it acts on x. By Property (c) of a group action (see Definition 2.1):

$$\begin{aligned}(g_k^{-1} g) \wedge x &= g_k^{-1} \wedge (g \wedge x) \\ &= g_k^{-1} \wedge x_k.\end{aligned}$$

Note that g_k is *unique* once we have chosen the g_i.

But since g_k moves x to x_k, it follows that g_k^{-1} moves x_k to x.

Thus,

$$(g_k^{-1} g) \wedge x = x,$$

so that $g_k^{-1} g$ is one of the elements of $\operatorname{Stab}(x)$, say

$$g_k^{-1} g = h_q. \tag{3.4}$$

Note that h_q is *unique* since g_k is.

Left-multiplying both sides of Equation 3.4 by g_k:

$$g_k g_k^{-1} g = g_k h_q;$$
$$g = g_k h_q.$$

Thus, g is the product of an element of $\text{Stab}(x)$ and one of the m particular elements $g_1, \ldots, g_m$. Moreover, the above argument shows that there is *exactly one* g_k and *exactly one* h_q such that $g = g_k h_q$. But as this is true of *each* element of G, it follows that

$$|G| = |\{g_1, \ldots, g_m\}| \times |\{h_1, \ldots, h_n\}|$$
$$= |\text{Orb}(x)| \times |\text{Stab}(x)|. \qquad \blacksquare$$

This has still not *quite* established Equation 3.1, but we are nearly there!

Corollary

When a finite group G acts on a finite set X, then

$$\text{number of orbits} = \frac{1}{|G|} \sum_{x \in X} |\text{Stab}(x)|.$$

Proof

Let us work out the sum

$$\sum_{x \in X} |\text{Stab}(x)|,$$

an orbit at a time! Suppose $a, b, \ldots, k$ are elements of X, *one from each orbit*; then

$$\sum_{x \in X} |\text{Stab}(x)| = \sum_{x \in \text{Orb}(a)} |\text{Stab}(x)| + \sum_{x \in \text{Orb}(b)} |\text{Stab}(x)| + \cdots$$
$$+ \sum_{x \in \text{Orb}(k)} |\text{Stab}(x)|. \qquad (3.5)$$

Now within a particular orbit, say $\text{Orb}(a)$, the Orbit–stabilizer Theorem shows that $|\text{Stab}(x)|$ must be a constant, since $|G|$ is constant and we are staying within the same orbit. Thus, by the time we have counted $|\text{Stab}(x)|$ over each of the $|\text{Orb}(a)|$ elements of $\text{Orb}(a)$, the total we get is $|G|$, by the Orbit–stabilizer Theorem. This is true for each orbit; so Equation 3.5 becomes

$$\sum_{x \in X} |\text{Stab}(x)| = |G| + |G| + \cdots + |G|$$
$$= (\text{number of orbits}) \times |G|.$$

Rearranging and dividing by $|G|$:

$$\text{number of orbits} = \frac{1}{|G|} \sum_{x \in X} |\text{Stab}(x)|.$$

This completes the proof of the corollary. $\qquad \blacksquare$

Equation 3.1 is now at last established!

4 THE CYCLE INDEX

There is much hard work involved in the use of the Counting Lemma to count equivalence classes. For larger problems, it may become impractical to determine explicitly which colourings are fixed by a particular element of the symmetry group. Can we find some way round this problem?

The first thing to notice is that actually we do not need to know *which* colourings are fixed, only *how many* of them are. We have to find a way of counting them *without* looking in detail at all of them.

You have already had some practice at doing this in Subsection 3.4. The solutions, as you may have noticed, did *not* explicitly draw up all the individual colourings; instead, they assessed the size of $|\text{Fix}(g)|$ for each g by *counting up how many independent colour choices could be made*. The development of this technique is the subject of this section.

4.1 Permutations and cycles

Let us take yet another look at our example of the colouring of the sides of a square picture frame. This time, however, we shall use three colours.

Exercise 4.1

Refer back to our treatment of the problem in Exercise 3.2(a) of the previous section, and adapt it to find the number of equivalence classes when *three* colours are available.

How do the solutions of Exercise 3.2(a) and Exercise 4.1 compare? The sizes of the fixed sets in the two cases are as follows:

	$|\text{Fix}(e)|$	$|\text{Fix}(r)|$	$|\text{Fix}(r^2)|$	$|\text{Fix}(r^3)|$
Two colours	2^4	2^1	2^2	2^1
Three colours	3^4	3^1	3^2	3^1

Table 4.1

Thus, the numbers of equivalence classes in the two cases were as follows:

Two colours: number of equivalence classes $= \frac{1}{4}(2^4 + 2^1 + 2^2 + 2^1) = 6$.

Three colours: number of equivalence classes $= \frac{1}{4}(3^4 + 3^1 + 3^2 + 3^1) = 24$.

There are no prizes for guessing the answer for 4 colours! In fact, for m colours the number of equivalence classes will be

$$\tfrac{1}{4}(m^4 + m^1 + m^2 + m^1) = \tfrac{1}{4}(m^4 + m^2 + 2m).$$

But what about the indices? Where do they come from?

Exercise 4.2

Recall the square picture frame, and remember that we labelled the positions of the sides as shown in Figure 4.1. For each of the direct symmetries, write down the permutation of the position labels it produces.

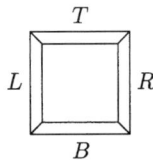

Figure 4.1

If you used cycle form, you probably gave the answer for the identity symmetry simply as (T) — or perhaps even as e. Writing it out in full, as $(T)(L)(B)(R)$, will reveal the pattern we are looking for.

Consider, for instance, the cycle form for r^2: $(TB)(LR)$. This notation shows that r^2 is the product of two disjoint *cycles of length 2*, or *2-cycles*, where a **cycle of length k** (or ***k*-cycle**) is a cycle involving exactly k positions. Our earlier argument was that opposite edges must have the same colour, i.e. that the edges at positions T and B must have the same colour and that those at positions L and R must have the same colour. But this says that labels in the same cycle refer to positions all of which must have the same colour.

Thus $e = (T)(L)(B)(R)$ is the product of four disjoint 1-cycles.

Consequently, the choice is of one colour for each cycle. If there are m colours, the total choice is thus m^2, since there are two cycles in the permutation. In general, we shall have

$$\text{Fix}(g) = m^{\#(g)},$$

where $\#(g)$ is the number of cycles in the permutation on the positions induced by g — with all 1-cycles being counted.

From now on, we shall refer simply to the 'number of cycles in a permutation' on the understanding that we mean *disjoint* cycles.

This allows us to state an important result, actually a special case of *Pólya's Theorem*, which we shall establish in the next section.

Theorem 4.1 Orbit Counting Theorem

If G is the group of symmetries of some configuration X which is to be coloured using m colours, then the total number of orbits of colourings is given by

$$\frac{1}{|G|} \sum_{g \in G} m^{\#(g)},$$

where $\#(g)$ is the number of cycles (including 1-cycles) in the cycle form of the permutation of the position labels induced by g.

Exercise 4.3

Apply the above Orbit Counting Theorem to the colourings of the four sides of a square picture frame with three colours, using the group $G = \{e, r, r^2, r^3\}$.

4.2 The cycle symbol

Although it is clear from Theorem 4.1 that the number of disjoint cycles in each permutation is all that is needed to enable us to count the number of equivalence classes of colourings, we shall need more information about the cycle structure of the permutations if we want to *identify* all the equivalence classes of colourings rather than just to *count* them.

To this end, we make use of a notation to record the cycle structures. This will take account, not of what each cycle does, but simply of its *length* — and how many cycles of each length there are. So, we shall denote a cycle of length k by the symbol x_k, and we shall represent any symmetry by a product with a factor x_k for *each* cycle of length k in the cycle form of the corresponding permutation.

Definition 4.1 Cycle symbol

Let g be a symmetry of a configuration X. Then the **cycle symbol** $cs(g)$ is an algebraic term consisting of a product of powers of variables x_k, where for each k the power of x_k in $cs(g)$ is the number of cycles of length k in the permutation induced by g on the position labels.

This definition is rather intimidating but a few examples should clarify its meaning.

Example 4.1

Take X to be our familiar square picture frame with side positions T, L, B, R. What is $cs(g)$ for each g in the rotation group $\{e, r, r^2, r^3\}$?

The cycle representation of e is $(T)(L)(B)(R)$ — four 1-cycles. Thus in the cycle symbol, we must raise x_1 to the fourth power:

$$cs(e) = x_1^4.$$

The cycle representation of r is $(TLBR)$ — one 4-cycle. Thus in the cycle symbol, we must raise x_4 to the first power:

$$cs(r) = x_4.$$

The cycle representation of r^2 is $(TB)(LR)$ — two 2-cycles. Thus in the cycle symbol, we must raise x_2 to the second power:

$$cs(r^2) = x_2^2.$$

Finally, it is clear that r^3 has the same cycle symbol as r:

$$cs(r^3) = x_4. \qquad \blacklozenge$$

Example 4.2

Take X to be a hexagon split into three rhombuses, with symmetry group $D_3 = \{e, r, r^2, s, rs, r^2s\}$, where $r = r[2\pi/3]$ and $s = q[0]$. The symmetry group acts on the three rhombus positions $1, 2, 3$ (see Figure 4.2). What is $cs(g)$ for each $g \in D_3$?

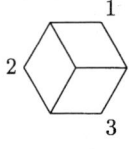

Figure 4.2

The cycle representation of e in this case is $(1)(2)(3)$ — three 1-cycles. Thus,

$$cs(e) = x_1^3.$$

The cycle representation of r is (123) — one 3-cycle. Similarly, r^2 gives one 3-cycle. Thus,

$$cs(r) = cs(r^2) = x_3.$$

The cycle representation of s is $(2)(13)$ — a 1-cycle and a 2-cycle. The elements rs and r^2s have similar cycle structures. Thus,

$$cs(s) = cs(rs) = cs(r^2s) = x_1 x_2. \qquad \blacklozenge$$

Exercise 4.4

Let X be the square picture frame with side positions T, L, B, R, with symmetry group $D_4 = \{e, r, r^2, r^3, s, rs, r^2s, r^3s\}$. Find the cycle symbol of each indirect symmetry.

Exercise 4.5

Let X be the square picture frame, this time with five positions. T, L, B and R are as before, but the central square hole where the picture goes is counted as a fifth position, C (see Figure 4.3). Find $cs(g)$ for each $g \in D_4$.

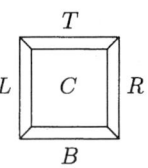

Figure 4.3

Now the practical purpose of cycle symbols is to make it easier to use the Counting Lemma, which always involves summing over all the group elements then dividing by $|G|$. Thus, it should not surprise you to learn that there is a name for the result of doing this.

> **Definition 4.2 Cycle index**
>
> Let G be a symmetry group acting on a configuration X. When G acts in this way, its **cycle index** is given by
>
> $$P_G(x_1, x_2, \ldots) = \frac{1}{|G|} \sum_{g \in G} cs(g).$$

The cycle index is denoted by $P_G(x_1, x_2, \ldots)$ to emphasize that it is a polynomial in the variables $x_1, x_2, \ldots$. We shall soon see that the real power of this polynomial is revealed when particular quantities are substituted for those variables.

Example 4.3

Write down the cycle index of G for each of the cases in Examples 4.1 and 4.2.

For Example 4.1, we have

$$\begin{aligned} P_G(x_1, x_2, x_3, x_4) &= \tfrac{1}{4}\left(cs(e) + cs(r) + cs(r^2) + cs(r^3)\right) \\ &= \tfrac{1}{4}\left(x_1^4 + x_4 + x_2^2 + x_4\right) \\ &= \tfrac{1}{4}\left(x_1^4 + x_2^2 + 2x_4\right). \end{aligned}$$

For Example 4.2, we have

$$\begin{aligned} P_G(x_1, x_2, x_3) &= \tfrac{1}{6} \sum_{g \in G} cs(g) \\ &= \tfrac{1}{6}\left(x_1^3 + 2x_3 + 3x_1 x_2\right). \end{aligned}$$ ♦

Exercise 4.6

Calculate the cycle index of the symmetry group D_4 for the case of the square picture frame as in Exercise 4.4.

Exercise 4.7

Calculate the cycle index of the symmetry group D_4 for the case of the square picture frame as in Exercise 4.5.

These cycle indices may look like pieces of pure algebraic mystique, but in the next subsection you will see how to set them to work.

4.3 Counting colourings

In Example 4.3 we found that the cycle index of the rotation group, acting on the four positions T, L, B and R of a square picture frame, is

$$P_G(x_1, x_2, x_3, x_4) = \tfrac{1}{4}\left(x_1^4 + x_2^2 + 2x_4\right).$$

Now you may remember that in Subsection 4.1 we discovered a formula for the number of equivalence classes of colourings of such a frame (assuming that it cannot be turned over), if m colours are available:

number of equivalence classes $= \tfrac{1}{4}\left(m^4 + m^2 + 2m\right).$

That is to say, if we replace each of the somewhat mysterious *variables* x_1, x_2, etc. by the *constant* m, we obtain the answer to the corresponding counting problem, using m colours!

Exercise 4.8

Find a formula for the number of equivalence classes of colourings (using m colours) of a hexagon divided into three rhombuses, made of transparent material so that it can be turned over. Verify that this agrees with what you have discovered already for $m = 2, 3$.

This leads us to the following result.

> **Theorem 4.2 Cycle Index Theorem**
>
> If C is a set of colourings of a configuration with m colours, and if that configuration is acted upon by the permutation group G, then the number of equivalence classes of colourings in C is given by $P_G(m, m, m, \ldots)$.

Proof

By definition,
$$P_G(x_1, x_2, x_3, \ldots) = \frac{1}{|G|} \sum_{g \in G} cs(g)$$
and, if we remember that the cycle symbol $cs(g)$ is just a product of factors such as x_k, one for each of the $\#(g)$ cycles in the cycle decomposition of g, we can see that replacing every factor in $cs(g)$ by m will give us $m^{\#(g)}$. Hence
$$P_G(m, m, m, \ldots) = \frac{1}{|G|} \sum_{g \in G} m^{\#(g)},$$
and the Orbit Counting Theorem tells us that the right-hand side is equal to the number of equivalence classes of colourings. ■

Theorem 4.1

Exercise 4.9

Find a formula for the number of equivalence classes of colourings of a rectangular picture frame (cf. Exercise 3.4), using m colours, assuming that each side is the same colour on both surfaces and that the frame can be turned over.

Exercise 4.10

Find a formula for the number of equivalence classes of colourings of a rectangle divided into quadrants (cf. Exercise 3.5), assuming that it is made of transparent material and can be turned over.

5 PÓLYA'S ENUMERATION FORMULA

We are now ready to tackle the project which we mooted in the last section: to list colourings rather than just to count them. You will not be surprised to learn that once more we turn to the square picture frame to illustrate the idea.

5.1 Listing the colourings

First, we must develop some algebraic notation for the colourings. We proceed in much the same way as we did in developing the cycle index notation for a group in Section 4. There we *multiplied* the variables x_k together to produce the cycle symbol for an *individual* group element, then *added* the resulting expressions together to produce the cycle index representing the *group*. Now, we first *multiply* some terms together to represent an *individual* colouring, then we *add* the results to represent a *set* of colourings.

In the case of the cycle index, we also divided by the order of the group. At *this* stage in the process of listing the colourings, we are considering them individually, so we do *not* divide by $|G|$.

We begin by inventing symbols for the various colours. There is no standard terminology here, so let us be as straightforward as possible and use B for black, W for white, etc. Why complicate matters?

Example 5.1

Consider the colouring of the picture frame in which all four sides are black. By analogy with our construction of the cycle symbol, we represent this colouring algebraically by raising B to the fourth power: B^4. ♦

Exercise 5.1

Write down the symbols for:

(a) the colouring in which the top side is white and the others are black;

(b) the colouring in which the left and bottom sides are white and the other two are black.

The next step is to represent *sets* of colourings by *adding* the symbols for individual colourings.

Example 5.2

Consider the set of all colourings in which *at most one side is white*. There are five such colourings — one with all four sides black, one with just the top white, one with just the left side white, and so on. The first of these colourings has the symbol B^4, and *each* of the others has the symbol B^3W (or WB^3 — we count these as equivalent); so the algebraic symbol for the *set* is

$$B^4 + 4B^3W.$$ ♦

One can interpret this notation as a system in which *multiplication* means '*and*' while *addition* means '*or*'. Thus,

$$B^4 + 4B^3W$$

means 'the set in which *either* the colours are (black *and* black *and* black *and* black) *or* one of the *four* possibilities with (black *and* black *and* black *and* white) occurs'.

Exercise 5.2

Write down the symbols for:

(a) the set of colourings with two white and two black sides;

(b) the set of colourings in which at least three sides are white.

Exercise 5.3

Write down the symbols for the set of *all* colourings of the frame with the two colours black and white.

If you remember the Binomial Theorem, you may have noticed that the list of colourings just obtained is the expansion of the binomial $(B + W)^4$. These are the colourings left unchanged by the identity, since it leaves all colourings invariant.

The logic of the situation is that each side can be coloured black or white independently of the others. We thus get the binomial $B + W$ for each side — so the result is

$(B$ or $W)$ and $(B$ or $W)$ and $(B$ or $W)$ and $(B$ or $W)$.

If we denote 'or' by $+$ and 'and' by a product, we get the binomial $(B + W)^4$.

Exercise 5.4

Use the above notation to find the expression for the set of all possible colourings of the three rhombuses depicted in Figure 5.1 with the *three* colours black (B), grey (G) and white (W). Verify that the expression you obtain is in fact the cube of a simpler expression.

Figure 5.1

We shall call an expression of this kind a *colouring inventory*.

> ***Definition 5.1 Colouring inventory***
>
> Suppose we have a configuration X and a set C of possible colourings of X using some set of colours, to each of which is associated a label (for example, we have just used the initial letters B, G, W as labels). The **colouring inventory** of C is a polynomial in the labels, in which:
>
> (a) each term describes those colourings in C with some particular colour distribution; that is to say, the power of a particular label (say B) describes the number of positions that have that colour (so that B^2, for example, indicates two black positions);
>
> (b) the coefficient of a particular term describes the *number* of individual colourings in C with that colour distribution.

Exercise 5.5

Given the configuration in Exercise 5.4 and the colours black, grey and white, write down the colouring inventory of the set of colourings in which *at least two* rhombuses are coloured grey.

We saw in Exercise 5.3 and the discussion following it, that the set of *all* colourings of a square picture frame using black and white is described by the colouring inventory $(B + W)^4$. This set of colourings is, of course, Fix(e). Next, what about the colourings belonging to Fix(r) — those that are unchanged by rotation through $\pi/2$? We must have four black sides *or* four white sides — so the colouring inventory is $B^4 + W^4$.

By, continuing in this way, the inventories of the sets of fixed colourings for each of the group elements e, r, r^2, r^3 are given by

$$\begin{aligned}\text{Fix}(e) &: B^4 + 4B^3W + 6B^2W^2 + 4BW^3 + W^4 = (B^1+W^1)^4 \\ \text{Fix}(r) &: B^4 + W^4 = (B^4+W^4)^1 \\ \text{Fix}(r^2) &: B^4 + 2B^2W^2 + W^4 = (B^2+W^2)^2 \\ \text{Fix}(r^3) &: B^4 + W^4 = (B^4+W^4)^1.\end{aligned}$$

Now in Section 3 we discovered that we obtain the correct *number* of equivalence classes of colourings by adding |Fix(g)| for each g, then dividing by the order of the group. We now have a colouring inventory for Fix(g) for each g; what happens if we add *these* expressions together and divide by |G|? In this case, we obtain

$$\tfrac{1}{4}\left[(B^1+W^1)^4 + (B^4+W^4)^1 + (B^2+W^2)^2 + (B^4+W^4)^1\right]$$

$$\begin{aligned}&= \tfrac{1}{4}(B^4 + 4B^3W + 6B^2W^2 + 4BW^3 + W^4 \\ &\phantom{=\tfrac{1}{4}(}+ B^4 + W^4 \\ &\phantom{=\tfrac{1}{4}(}+ B^4 + 2B^2W^2 + W^4 \\ &\phantom{=\tfrac{1}{4}(}+ B^4 + W^4) \\ &= \tfrac{1}{4}(4B^4 + 4B^3W + 8B^2W^2 + 4BW^3 + 4W^4) \\ &= \phantom{\tfrac{1}{4}(4}B^4 + B^3W + 2B^2W^2 + BW^3 + W^4\end{aligned}$$

and the terms in the expression now correspond exactly to the six *equivalence classes* of colourings! To the term $2B^2W^2$ there correspond two distinct equivalence classes: one with two adjacent black sides and two adjacent white sides, the other with alternate black and white sides.

Exercise 5.6

Consider the rectangular picture frame with symmetry group
$\Gamma(\Box) = \{e, r, h, v\}$ (see Exercise 3.4). Write down the colouring inventory of
$\text{Fix}(g)$ for each $g \in \Gamma(\Box)$, and verify that taking the sum and dividing by 4
results in a list of the equivalence classes of colourings using the colours
black and white.

In order to distinguish between lists of *individual* colourings and lists of
equivalence classes of colourings, we have a different name for the latter.

> *Definition 5.2 Pattern inventory*
>
> The **pattern inventory** of a set of equivalence classes of colourings of
> a configuration is a polynomial which lists these classes in the same
> way that a colouring inventory lists individual colourings.

Example 5.3

In the case of the square picture frame with black or white sides which
cannot be turned over, we saw, just before Exercise 5.6, that the terms in
the expression

$$B^4 + B^3W + 2B^2W^2 + BW^3 + W^4$$

correspond to the six equivalence classes of colourings. (Note that there are
two equivalence classes in which two sides are coloured black and two
white.) This expression is therefore the pattern inventory for this set of
equivalence classes of colourings. ♦

Exercise 5.7

Take X to be a hexagon divided into three rhombuses, with symmetry
group D_3, as in Example 4.2. Assuming that the three colours black, grey
and white are available, write out the pattern inventory.

Do you feel 'all dressed up and nowhere to go'? You now have an algebraic
notation for *describing* colourings and their equivalence classes — but you
probably did the last exercise by using common sense, first of all, to find the
equivalence classes and then translating this into the algebraic notation.
You did not actually *use* the notation to solve the problem.

Despair not — you have not laboured in vain! We now have all the notation
and concepts at our disposal to state and prove Pólya's Theorem — which
describes how to generate the pattern inventory simply by making
substitutions in the cycle index. And this works for a configuration of any
degree of complexity, using any number of colours!

5.2 Pólya's Theorem

Before proceeding to the theorem itself, let us look back at some cycle
symbols and compare them with the corresponding colouring inventories.

Example 5.4

In Example 4.1, we discovered that for the group of rotational symmetries
acting on the square picture frame, the cycle symbols are

$$cs(e) = x_1^4, \quad cs(r) = x_4, \quad cs(r^2) = x_2^2, \quad cs(r^3) = x_4,$$

and later we discovered the colouring inventories for $\text{Fix}(e)$, $\text{Fix}(r)$, $\text{Fix}(r^2)$
and $\text{Fix}(r^3)$ for this configuration. They are, respectively,

$$(B+W)^4, \quad B^4 + W^4, \quad (B^2 + W^2)^2, \quad B^4 + W^4.$$

Can you see the connection?

Next, let us look at the solutions to Exercise 4.9 and Exercise 5.6, which involved the rectangular picture frame. From the solution to Exercise 4.9 we see that the cycle symbols are

$$cs(e) = x_1^4, \quad cs(r) = x_2^2, \quad cs(h) = x_1^2 x_2, \quad cs(v) = x_1^2 x_2.$$

The colouring inventories in the solution to Exercise 5.6, on the other hand, can be written respectively as

$$(B+W)^4, \quad (B^2+W^2)^2, \quad (B+W)^2(B^2+W^2), \quad (B+W)^2(B^2+W^2).$$

(Check this for yourself!)

The distinct cycle symbols and colouring inventories which occur in these two configurations are listed in Table 5.1.

Cycle symbol	Colouring inventory
x_1^4	$(B+W)^4$
x_4	$B^4 + W^4$
x_2^2	$(B^2+W^2)^2$
$x_1^2 x_2$	$(B+W)^2(B^2+W^2)$

Table 5.1

Before reading on, please try to see for yourself the connection between the two columns of this table.

The connection is this: each occurrence of x_1 in the cycle symbol becomes $B+W$ in the colouring inventory (so that x_1^4 becomes $(B+W)^4$); each occurrence of x_2 becomes B^2+W^2; and, in general, each occurrence of x_i becomes $B^i + W^i$. ♦

In fact, it is quite easy to prove the following lemma.

Lemma 5.1

Let X be any configuration, acted upon by any symmetry group G; let $\{p_1, \ldots, p_n\}$ be a set of labels for n colours; and let C be the set of all colourings of X using these colours. Then for each $g \in G$, the colouring inventory for Fix(g) is obtained from the cycle symbol $cs(g)$ by substituting $p_1^i + \cdots + p_n^i$ for x_i, for each relevant i.

Proof

Suppose that there are N positions in the configuration and that the cycle representation of g contains just one cycle, $(a_1 a_2 \ldots a_N)$. Then Fix(g) must consist of colourings in which all N positions have the same colour, which may be p_1 or p_2 or ... or p_n. Thus, the colouring inventory in this case is $p_1^N + p_2^N + \cdots + p_n^N$ and, since $cs(g) = x_N$, the statement of the lemma is true in this case.

Suppose next that the cycle representation of g contains several cycles. Then we can consider X to be made up of several sub-configurations, each corresponding to a cycle of g. Each can be coloured independently of the others, so the inventory for the colourings of X is the product of the inventories for each cycle separately. Since a cycle of length i has the colouring inventory $p_1^i + \cdots + p_n^i$, this proves the result. ∎

Example 5.5

Consider (yet again!) the square picture frame, with three possible colours: red, yellow and green, just to brighten things up. Thus, our labels can be $p_1 = R, p_2 = Y, p_3 = G$. Consider the group element $g = s$, and let us compute the colouring inventory for Fix(s).

From the solution to Exercise 4.4, the cycle symbol is

$$cs(s) = x_2 x_1^2.$$

Thus, according to Lemma 5.1, we must replace

$$x_2 \quad \text{by} \quad p_1^2 + p_2^2 + p_3^2 = R^2 + Y^2 + G^2$$

and

$$x_1 \quad \text{by} \quad p_1 + p_2 + p_3 = R + Y + G.$$

Thus, from $x_2 x_1^2$ we derive the following colouring inventory

$$\left(R^2 + Y^2 + G^2\right)(R + Y + G)^2. \qquad \blacklozenge$$

Without multiplying this out, we can use it to illustrate the reasoning in the proof of Lemma 5.1. The effect of s is to interchange the top and bottom sides, so they give rise to two red *or* two yellow *or* two green sides. This gives the expression $\left(R^2 + Y^2 + G^2\right)$. Quite independently of this, the left side can be coloured red *or* yellow *or* green $(R + Y + G)$, and the right side can be coloured red *or* yellow *or* green $(R + Y + G)$. These choices are independent, so we multiply to obtain the final result

$$\left(R^2 + Y^2 + G^2\right)(R + Y + G)^2.$$

Exercise 5.8

If the hexagon in Figure 5.1 is coloured using a selection from red, yellow, green and blue, what is the colouring inventory of Fix(s), where once again s is reflection in the horizontal axis?

We are now in a position to state and prove Pólya's Theorem which asserts that the pattern inventory is obtained from the cycle index in exactly the same way that one obtains the colouring inventory from the corresponding cycle symbol!

> *Theorem 5.1 Pólya's Theorem*
>
> Let X be any configuration, acted upon by any symmetry group G, and let $\{p_1, \ldots, p_n\}$ be a set of labels for n colours. Then the pattern inventory for the equivalence classes of these colourings is obtained from the cycle index P_G by substituting $p_1^i + \cdots + p_n^i$ for x_i, for each relevant i. That is to say, the pattern inventory is
>
> $$P_G\left(\sum p_j, \sum p_j^2, \ldots, \sum p_j^n\right).$$

Proof

For any particular colour distribution, the number of equivalence classes of colourings with that distribution is obtained by summing over the numbers of colourings with that distribution belonging to Fix(g), and dividing by $|G|$; this is simply the Counting Lemma applied to the colourings with that distribution. As this is true of all colour distributions, it follows that summing the colouring inventories of Fix(g) and dividing by $|G|$ gives the pattern inventory. But by Lemma 5.1, each colouring inventory is obtained by making the stated substitution in $cs(g)$. Therefore, if we make the stated substitution in $\dfrac{1}{|G|} \sum_{g \in G} cs(g)$ (which is $P_G(x_1, \ldots, x_n)$), we obtain the pattern inventory. $\blacksquare$

Exercise 5.9

Let X be a rectangular picture frame, with symmetry group $\Gamma(\square) = \{e, r, h, v\}$ (i.e. it can be turned over), where each side is the same colour (black, grey or white) on both surfaces. Find the pattern inventory. (Do not expand it.)

Exercise 5.10

Let X be a rectangle divided into quadrants, made of transparent material, with symmetry group $\Gamma(\square) = \{e, r, h, v\}$. Find the pattern inventory, again for the three colours black, grey, white. (Do not expand it.)

Exercise 5.11

A Star of David brooch is made from two equilateral triangles, each with three sides. Each of these sides independently is made of silver or gold. The two triangles are welded together, so that the symmetry group of the positions of the edges is D_3 (see Figure 5.2). Find the pattern inventory, and the number of equivalence classes.

Figure 5.2

If a customer can afford at most three gold sides, how many equivalence classes can she choose from?

We hope you can appreciate the enormous flexibility of Pólya's Theorem, and agree that it makes a fitting culmination to our study of counting with groups. Why not make up (and solve) some counting problems of your own?

APPENDIX: COLOURING THE CUBE

The material in this appendix is optional.

We promised in the Introduction that we would return to the problem of colouring the faces of a cube using three colours. We shall in fact do more than this; we shall use the Cycle Index Theorem to obtain a general formula for the number of equivalence classes using m colours, and we shall derive the pattern inventory for the three colours red (R), white (W), blue (B).

As always, we have the choice between considering the group of allowable transformations to be the full symmetry group (including reflections) or the direct symmetry group. Using the full symmetry group would imply regarding a colouring and its mirror image as being in the same equivalence class but, clearly, if you put a number of coloured cubes into a bag and shake them, none of them will undergo reflection — we would have to push a cube into a fourth or higher dimension, rotate it and bring it back into three dimensions in order to achieve that! So we shall regard G as the rotation group of the cube:

$$G = \Gamma^+(\text{cube}).$$

What are the rotational symmetries of the cube? We shall cover this again in *Unit GE5*, but for the moment let us observe that there are three kinds of axis of symmetry: through pairs of opposite faces, through the midpoints of opposite sides and through pairs of opposite vertices (see Figure A.1).

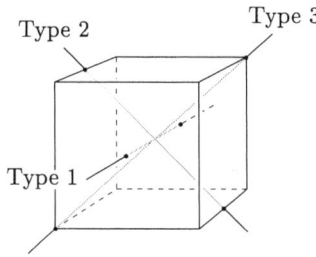

Figure A.1

Let us now label the positions of the faces of the cube $1, \ldots, 6$ (see Figure A.2).

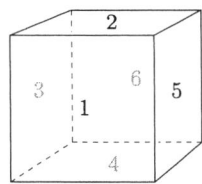

Figure A.2

The positions of the three visible faces are 1, 2, and 5; the opposite hidden faces are 6, 4 and 3 respectively.

We now have to find the cycle symbol of each element of G. The identity is easy: its cycle representation is $(1)(2)(3)(4)(5)(6)$, so that

$$cs(e) = x_1^6.$$

Now let r be a rotation through $\pi/2$ about the Type 1 axis shown in Figure A.1, anticlockwise when viewed facing position 1. Its cycle representation is $(1)(6)(2345)$, so

$$cs(r) = x_1^2 x_4.$$

Each of the three Type 1 axes has two rotations of this type, so there are six elements of G with this cycle symbol.

Next, the cycle symbol for r^2 is $(1)(6)(24)(35)$, so

$$cs(r^2) = x_1^2 x_2^2.$$

Each Type 1 axis has one rotation of this type, so there are three elements of G with this cycle symbol.

Consider now the Type 2 axes. Each rotation about such an axis has a cycle representation such as $(16)(23)(45)$ and cycle symbol x_2^3; there are six such rotations.

Finally, each rotation about a Type 3 axis has a cycle representation such as $(143)(256)$ and cycle symbol x_3^2; there are eight such rotations.

Adding all this up,
$$P_G(x_1, x_2, x_3, x_4, x_5, x_6) = \tfrac{1}{24}\left(x_1^6 + 6x_1^2 x_4 + 3x_1^2 x_2^2 + 6x_2^3 + 8x_3^2\right).$$

Using the Cycle Index Theorem, the number of equivalence classes of colourings using m colours is obtained by substituting m for each of the variables $x_1, \ldots, x_6$:

$$\text{number of equivalence classes} = \tfrac{1}{24}\left(m^6 + 6m^3 + 3m^4 + 6m^3 + 8m^2\right)$$
$$= \tfrac{1}{24}\left(m^6 + 3m^4 + 12m^3 + 8m^2\right).$$

For the particular case of three colours, this gives 57 varieties!

By Pólya's Theorem, the pattern inventory for R, W and B is given by substituting $R^i + W^i + B^i$ for x_i ($i = 1, \ldots, 6$), to obtain

$$\tfrac{1}{24}\big((R+W+B)^6 + 6(R+W+B)^2(R^4+W^4+B^4)$$
$$+ 3(R+W+B)^2(R^2+W^2+B^2)^2$$
$$+ 6(R^2+W^2+B^2)^3 + 8(R^3+W^3+B^3)^2\big).$$

It would be possible to go on to expand this polynomial and thus list the number of equivalence classes of colourings with every possible colour distribution. Suppose, however, that you merely want to know how many equivalence classes with two red, two white and two blue faces there are. Then your work can be considerably shortened because the answer is simply the coefficient of $R^2 W^2 B^2$ in the expansion. Now:

the coefficient of $R^2 W^2 B^2$ in $(R+W+B)^6$ is $\dfrac{6!}{2!2!2!} = 90$;

the coefficient of $R^2 W^2 B^2$ in $6(R+W+B)^2(R^4+W^4+B^4)$ is 0;

that in $3(R+W+B)^2(R^2+W^2+B^2)^2$ is 18;

that in $6(R^2+W^2+B^2)^3$ is 36;

and that in $8(R^3+W^3+B^3)^2$ is 0.

Thus the coefficient of $R^2 W^2 B^2$ in the whole pattern inventory is
$$\tfrac{1}{24}(90 + 0 + 18 + 36 + 0) = 6.$$

SOLUTIONS TO THE EXERCISES

Solution 1.1

Four, represented by frames 1, 2, 4 and 6.

Frame 1 has two adjacent sides black and two white, and the same is true of 3, 5 and 9. Frame 2 has two opposite sides black and two white, and the same is true of 7. Frame 4 has three black sides, and the same is true of 8. Frame 6 has three white sides, and this is not true of any other frame in the figure.

Solution 1.2

This time, there are eight individual colourings, divided into four classes of equivalent colourings as follows.

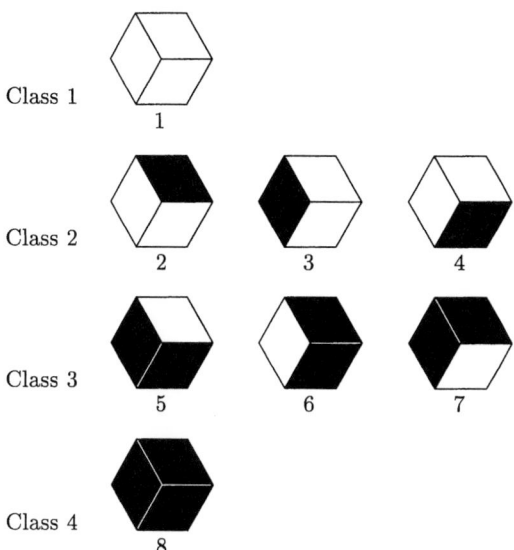

Solution 1.3

(a) Applying e, $r[2\pi/3]$ and $r[4\pi/3]$ gives the following.

(b) In addition to the above, applying $q[0]$, $q[\pi/3]$ and $q[2\pi/3]$ gives the following.

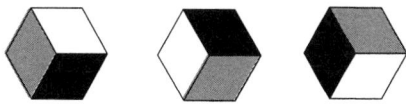

Solution 2.1

	A	B	C	D
e	A	B	C	D
r	B	C	D	A
r^2	C	D	A	B
r^3	D	A	B	C
s	D	C	B	A
rs	A	D	C	B
r^2s	B	A	D	C
r^3s	C	B	A	D

Solution 2.2

	AC	BD
e	AC	BD
r	BD	AC
r^2	AC	BD
r^3	BD	AC
s	BD	AC
rs	AC	BD
$r^2 s$	BD	AC
$r^3 s$	AC	BD

Solution 2.3

	c_1	c_2	c_3	c_4	c_5	c_6	c_7	c_8
e	c_1	c_2	c_3	c_4	c_5	c_6	c_7	c_8
r	c_1	c_3	c_4	c_2	c_6	c_7	c_5	c_8
r^2	c_1	c_4	c_2	c_3	c_7	c_5	c_6	c_8

The orbits are $\{c_1\}$, $\{c_2, c_3, c_4\}$, $\{c_5, c_6, c_7\}$ and $\{c_8\}$.

Solution 2.4

This is the same group action table as for the corners of the square, and is given as Solution 2.1.

Solution 2.5

The six colourings are as follows.

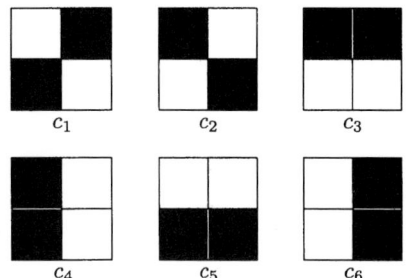

The group action table is thus:

	c_1	c_2	c_3	c_4	c_5	c_6
e	c_1	c_2	c_3	c_4	c_5	c_6
r	c_2	c_1	c_4	c_5	c_6	c_3
r^2	c_1	c_2	c_5	c_6	c_3	c_4
r^3	c_2	c_1	c_6	c_3	c_4	c_5
s	c_2	c_1	c_5	c_4	c_3	c_6
rs	c_1	c_2	c_6	c_5	c_4	c_3
$r^2 s$	c_2	c_1	c_3	c_6	c_5	c_4
$r^3 s$	c_1	c_2	c_4	c_3	c_6	c_5

Under the action of the group, there are two orbits:

$$\{c_1, c_2\} \quad \text{and} \quad \{c_3, c_4, c_5, c_6\}.$$

The first orbit consists of those colourings in which one pair of diagonally opposite quadrants is one colour and the other pair is the other colour. The second orbit consists of those colourings in which an adjacent pair is one colour and the other adjacent pair is the other colour.

Solution 3.1

We could just add together the answers to Examples 3.1 and 3.2 to get $9 + 12 = 21$ equivalence classes (orbits).

Alternatively, we could count all 'effective objects' again, counting every object with a symmetry group of order 2 twice. This gives a total of $36 + 48 = 84$ 'effective objects', which divides by 4 to get 21 equivalence classes (orbits).

Solution 3.2

(a) As the frame cannot be turned over, the group of allowable transformations is $\{e, r, r^2, r^3\}$ again. For each group element, we have to find how many colourings it does not change.

e:	all 2^4 colourings are left unchanged, so that is the size of Fix(e):	16
r:	each side must have the same colour as the next, so all 4 sides are the same colour, with just two choices:	2
r^2:	opposite sides must be the same colour, so there are two pairs to colour independently, in 2^2 ways:	4
r^3:	each side must have the same colour as the one on its left, so all four are the same colour, with just two choices:	2

From this we calculate the sum

$$\sum_{g \in G} |\text{Fix}(g)| = 16 + 2 + 4 + 2 = 24,$$

and then, dividing by the order of the group, we get the number of orbits as

$$\frac{1}{|G|} \sum_{g \in G} |\text{Fix}(g)| = \tfrac{1}{4}(24) = 6,$$

which is, of course, exactly what we got in Section 1.

(b) The group of allowable transformations is now $\{e, r, r^2, r^3, s, rs, r^2s, r^3s\}$. We have calculated $|\text{Fix}(g)|$ for the first four of these elements already, in part (a). The other four elements are reflections, in the lines through the midpoints of the sides and the corners. In the figure, each of these axes is labelled with the corresponding group element. Again, we must discover how many colourings each reflection leaves unchanged.

s:	the sides in positions T and B must be the same colour, but those in positions L and R may be coloured independently, giving 2^3 choices:	8
rs:	the sides in positions T and R must be the same colour, as must be those in positions L and B, giving 2^2 choices:	4
r^2s:	a similar situation to s:	8
r^3s:	a similar situation to rs:	4

Adding these to the total of 24 from the direct symmetries, we now have

$$\sum_{g \in G} |\text{Fix}(g)| = 24 + 8 + 4 + 8 + 4 = 48.$$

But the order of the group is now 8, so we have

$$\frac{1}{|G|} \sum_{g \in G} |\text{Fix}(g)| = 48/8 = 6,$$

— which is exactly the same as we got before!

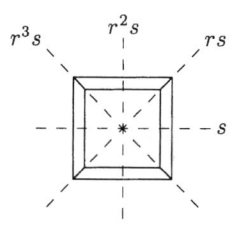

You could establish this using the Isometry Toolkit, but as long as you identify all four reflections in some form, this is all that matters.

The reason for this result is that, although the symmetry group has been enlarged, the equivalence classes remain the same as they were before. With only two colours, if two colourings are equivalent by a direct symmetry of the frame, then they are also equivalent by an indirect symmetry, and vice versa.

Solution 3.3

(a) Apart from the fact that sides become quadrants, the argument is exactly as in the solution to Exercise 3.2(a), giving again 6 equivalence classes.

(b) The argument here is very similar to the solution to Exercise 3.2(b), but it is worth noting that the reflections s, r^2s interchange roles with the reflections rs, r^3s.

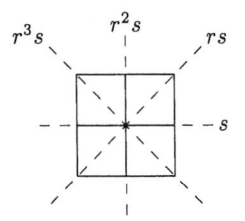

Here we have
$$|\text{Fix}(s)| = |\text{Fix}(r^2s)| = 4,$$
$$|\text{Fix}(rs)| = |\text{Fix}(r^3s)| = 8,$$

since in this case it is s and r^2s that interchange quadrants in pairs (in the same way that rs and r^3s interchanged sides in pairs in the case of the picture frame). Conversely, here it is rs and r^3s that each leave two quadrants in the same position while interchanging the other two.

Again, therefore, there are 6 equivalence classes.

Solution 3.4

(a) The group here is just $\{e, r\}$ where in this case $r = r[\pi]$. As before, we calculate how many colourings each does not change.

e: all 2^4 colourings are left unchanged: 16
r: the short sides must be the same colour, as must the long sides, giving 2^2 choices: 4

Thus the number of equivalence classes is
$$\frac{1}{|G|} \sum_{g \in G} |\text{Fix}(g)| = \tfrac{1}{2}(16 + 4) = 10.$$

(b) The group is the group of symmetries of the rectangle,
$\Gamma(\Box) = \{e, h, v, r\}$.

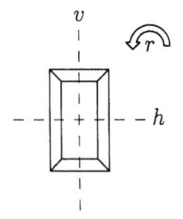

We calculated $|\text{Fix}(e)|$ and $|\text{Fix}(r)|$ in part (a). For the other two:

h: the short sides must be the same colour, giving 2^3 choices: 8
v: the long sides must be the same colour, giving 2^3 choices: 8

Thus the number of equivalence classes is
$$\tfrac{1}{4}(20 + 8 + 8) = 9.$$

Solution 3.5

(a) The group is as in the solution to Exercise 3.4(a).

e:	all 2^4 colourings are left unchanged:	16
r:	the top right and bottom left quadrants must be the same colour, as must the other two, giving 2^2 choices:	4

Thus the number of equivalence classes is

$$\frac{1}{|G|}\sum_{g \in G}|\text{Fix}(g)| = \tfrac{1}{2}(16+4) = 10.$$

(b) The group is again the group of symmetries of the rectangle, $\Gamma(\square) = \{e, h, v, r\}$.

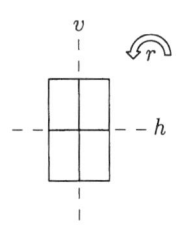

We calculated $|\text{Fix}(e)|$ and $|\text{Fix}(r)|$ in part (a). For the other two:

h:	the left quadrants must be the same colour, as must the right quadrants, giving 2^2 choices:	4
v:	the top quadrants must be the same colour, as must the bottom quadrants, giving 2^2 choices:	4

Thus the number of equivalence classes is

$$\frac{1}{|G|}\sum_{g \in G}|\text{Fix}(g)| = \tfrac{1}{4}(20+4+4) = 7.$$

Thus, rather surprisingly, the rectangular picture frame and the rectangle divided in quadrants behave the same way if they cannot be turned over, but differently if they can!

Solution 3.6

The group here is $\{e, r, r^2\}$. Working as in the last four exercises:

e:	all 3^3 colourings are left unchanged:	27
r:	each rhombus must have the same colour as the next, so all three are the same colour, with just three choices:	3
r^2:	same argument as for r:	3

Thus, the number of equivalence classes is

$$\frac{1}{|G|}\sum_{g \in G}|\text{Fix}(g)| = \tfrac{1}{3}(27+3+3) = 11.$$

Solution 3.7

The group is now $\{e, r, r^2, s, rs, r^3s\}$, where each reflection exchanges two rhombus positions and leaves the third fixed. We calculated $|\text{Fix}(e)|$, $|\text{Fix}(r)|$ and $|\text{Fix}(r^2)|$ in the solution to Exercise 3.6. As for the others:

s:	rhombus positions 1 and 3 must be the same colour, and the other rhombus can be coloured independently, giving 3^2 choices:	9
rs:	similar argument:	9
r^2s:	similar argument:	9

Thus, the number of equivalence classes is

$$\frac{1}{|G|}\sum_{g \in G}|\text{Fix}(g)| = \tfrac{1}{6}(33+9+9+9) = 10.$$

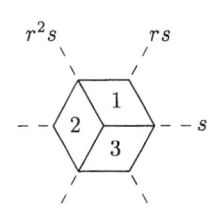

Solution 3.8

(a) The group is $\{e, r, r^2, r^3\}$ as for the square picture frame. As usual, we must find out, for each group element, how many objects it does not change.

e:	all 2^4 objects are left unchanged:	16
r:	if there is a particular orientation in the top right, then the orientation in the top left must be rotated through $\pi/2$ with respect to it, and similarly all the way round. Thus there are just two elements of Fix(r):	2

r^2:	the top right and bottom left must be the same orientation (since rotation of a 1 cm × 1 cm square through π leaves it looking the same). Similarly, the top left and bottom right must be in the same orientation, giving 2^2 choices:	4
r^3:	as r:	2

Thus the number of equivalence classes is

$$\frac{1}{|G|}\sum_{g \in G} |\text{Fix}(g)| = \tfrac{1}{4}(16 + 2 + 4 + 2) = 6.$$

(b) The group is $\{e, r, r^2, r^3, s, rs, r^2s, r^3s\}$, and we must now do the calculations for s, rs, r^2s and r^3s:

s:	the right pair must be in the same orientation, as must be the left pair, giving 2^2 choices:	4
rs:	Oh dear! The top right and bottom left squares must look the same after reflection on the diagonal passing through them. This is impossible!	0
r^2s:	the top pair must be in the same orientation, as must be the bottom pair, giving 2^2 choices:	4
r^3s:	impossible, for a similar reason to rs:	0

Thus the number of equivalence classes is

$$\frac{1}{|G|}\sum_{g \in G} |\text{Fix}(g)| = \tfrac{1}{8}(24 + 4 + 0 + 4 + 0) = 4.$$

Solution 3.9

This time the upper surface and the 'under' surface of each side of the frame must be regarded separately. Denoting the upper surfaces by T, L, B and R as before, we may denote the under surfaces by T', L', B' and R'.

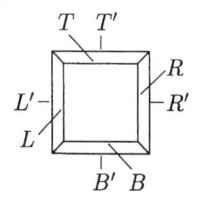

As before, we must now calculate how many colourings are not changed by each of the group elements e, r, r^2, r^3, s, rs, r^2s, r^3s.

e: all 2^8 colourings unchanged: 256
r: each of T, L, B, R must be the same colour, as must each
of T', L', B', R', giving 2^2 choices: 4
r^2: T and B must be the same colour, as must T' and B', L
and R, L' and R', giving 2^4 choices: 16
r^3: as for r: 4
s: L and L' must be the same colour, as must R and R', T
and B', T' and B, giving 2^4 choices: 16
rs: T and R' must be the same colour, as must T' and R, L
and B', L' and B, giving again 2^4 choices: 16
r^2s: similar situation to s: 16
r^3s: similar situation to rs: 16

Thus the total number of equivalence classes is

$$\frac{1}{|G|}\sum_{g\in G}|\text{Fix}(g)| = \tfrac{1}{8}(256+4+16+4+16+16+16+16) = \tfrac{1}{8}(344) = 43.$$

Solution 4.1

As before, for each element g of $\{e, r, r^2, r^3\}$ we must calculate $|\text{Fix}(g)|$.

e: all 3^4 colourings are left unchanged: 81
r: each side must have the same colour as the next, so all
four sides are the same colour, with just three choices: 3
r^2: opposite sides must be the same colour, so there are two
pairs to colour independently, in 3^2 ways: 9
r^3: each side must have the same colour as the one to its left,
so all four are the same colour, with just three choices: 3

From this we calculate the sum

$$\sum_{g\in G}|\text{Fix}(g)| = 81+3+9+3 = 96$$

and then, dividing by the order of the group, we get the number of equivalence classes as

$$\frac{1}{|G|}\sum_{g\in G}|\text{Fix}(g)| = \tfrac{1}{4}(96) = 24.$$

Solution 4.2

In two-line notation and cycle form, these permutations are as follows.

	two-line notation	cycle form
e	$\begin{pmatrix} T & L & B & R \\ T & L & B & R \end{pmatrix}$	$(T)(L)(B)(R)$
r	$\begin{pmatrix} T & L & B & R \\ L & B & R & T \end{pmatrix}$	$(TLBR)$
r^2	$\begin{pmatrix} T & L & B & R \\ B & R & T & L \end{pmatrix}$	$(TB)(LR)$
r^3	$\begin{pmatrix} T & L & B & R \\ R & T & L & B \end{pmatrix}$	$(TRBL)$

Solution 4.3

The numbers of cycles are as follows:

$$\#(e) = 4, \ \#(r) = 1, \ \#(r^2) = 2, \ \#(r^3) = 1,$$

and so the number of orbits (i.e. equivalence classes of colourings) is

$$\tfrac{1}{4}(3^4 + 3^1 + 3^2 + 3^1) = \tfrac{1}{4}(96) = 24,$$

in agreement with Solution 4.1.

Solution 4.4

The cycle representation of s is $(TB)(L)(R)$, and so
$$cs(s) = x_2 x_1^2.$$
The cycle representation of rs is $(TR)(LB)$, and so
$$cs(rs) = x_2^2.$$
Similarly,
$$cs(r^2 s) = x_2 x_1^2, \quad cs(r^3 s) = x_2^2.$$

Solution 4.5

The centre position always remains fixed. Thus, to each permutation, we must now adjoin the 1-cycle (C). Thus we must multiply by x_1 each cycle symbol from Example 4.1 and Exercise 4.4, to obtain
$$cs(e) = x_1^5,\ cs(r) = x_4 x_1,\ cs(r^2) = x_2^2 x_1,\ cs(r^3) = x_4 x_1,$$
$$cs(s) = x_2 x_1^3,\ cs(rs) = x_2^2 x_1,\ cs(r^2 s) = x_2 x_1^3,\ cs(r^3 s) = x_2^2 x_1.$$

Solution 4.6

By adding each cycle symbol from Example 4.1 and Exercise 4.4, and dividing by 8, the cycle index is given by
$$P_G(x_1, x_2, x_3, x_4) = \tfrac{1}{8}\left(x_1^4 + x_4 + x_2^2 + x_4 + x_2 x_1^2 + x_2^2 + x_2 x_1^2 + x_2^2\right)$$
$$= \tfrac{1}{8}\left(x_1^4 + 2x_1^2 x_2 + 3x_2^2 + 2x_4\right).$$

Solution 4.7

By adding each cycle symbol from Exercise 4.5, and dividing by 8, the cycle index is given by
$$P_G(x_1, x_2, x_3, x_4) = \tfrac{1}{8}\left(x_1^5 + 2x_1^3 x_2 + 3x_1 x_2^2 + 2x_1 x_4\right).$$

Solution 4.8

From Example 4.3, the cycle index is
$$P_G(x_1, x_2, x_3) = \tfrac{1}{6}\left(x_1^3 + 2x_3 + 3x_1 x_2\right),$$
and so, substituting m for each of x_1, x_2, x_3, we get $\tfrac{1}{6}(m^3 + 2m + 3m^2)$. In particular:

for two colours, $\tfrac{1}{6}(2^3 + 2 \times 2 + 3 \times 2^2) = 4$, which agrees with Solution 1.2;

for three colours, $\tfrac{1}{6}(3^3 + 2 \times 3 + 3 \times 3^2) = 10$, which agrees with Solution 3.7.

Note that Solution 1.2 did not actually consider the possibility of turning the hexagon over, but for two colours this makes no difference.

Solution 4.9

From the solution to Exercise 3.4, we have in this case
$$cs(e) = x_1^4,\ cs(r) = x_2^2,\ cs(h) = x_1^2 x_2,\ cs(v) = x_1^2 x_2,$$
so that
$$P_G(x_1, x_2) = \tfrac{1}{4}\left(x_1^4 + 2x_1^2 x_2 + x_2^2\right)$$
and the number of equivalence classes of colourings using m colours is
$$\tfrac{1}{4}\left(m^4 + 2m^3 + m^2\right).$$

Solution 4.10

From Solution 3.5, we have in this case
$$cs(e) = x_1^4, \ cs(r) = x_2^2, \ cs(h) = x_2^2, \ cs(v) = x_2^2,$$
so that
$$P_G(x_1, x_2) = \tfrac{1}{4}\left(x_1^4 + 3x_2^2\right),$$
and the number of equivalence classes of colourings using m colours is
$$\tfrac{1}{4}\left(m^4 + 3m^2\right).$$

Solution 5.1

(a) WB^3 (or B^3W).

(b) W^2B^2 (or B^2W^2).

Solution 5.2

(a) There are six individual colourings with two white and two black sides, so the symbol is $6W^2B^2$ (or $6B^2W^2$).

(b) $W^4 + 4W^3B$ (or $W^4 + 4BW^3$).

Solution 5.3

Just add the results for Example 5.2, Exercise 5.2(a) and Exercise 5.2(b), to obtain
$$B^4 + 4B^3W + 6B^2W^2 + 4BW^3 + W^4$$
(or any equivalent rearrangement).

Solution 5.4

There are 27 colourings, of which:

one is all black, giving B^3;

one is all grey, giving G^3;

one is all white, giving W^3;

three have two rhombuses black and one grey, giving $3B^2G$;

similarly, we obtain $3B^2W$, $3G^2B$, $3G^2W$, $3W^2B$ and $3W^2G$;

six have all three colours, giving $6BGW$.

Thus the expression is
$$B^3 + G^3 + W^3 + 3B^2G + 3B^2W + 3G^2B + 3G^2W + 3W^2B + 3W^2G + 6BGW.$$

It is easily checked (by multiplying out) that this is equal to $(B + G + W)^3$.

Solution 5.5

We just pick out, from the colouring inventory in Solution 5.4, those terms where G is raised to the power 2 or 3. Thus the requisite colouring inventory is $G^3 + 3G^2B + 3G^2W$.

Solution 5.6

The colouring inventories for the fixed sets are:
$$e : B^4 + 4B^3W + 6B^2W^2 + 4BW^3 + W^4$$
$$r : B^4 \qquad\quad + 2B^2W^2 \qquad\quad + W^4$$
$$h : B^4 + 2B^3W + 2B^2W^2 + 2BW^3 + W^4$$
$$v : B^4 + 2B^3W + 2B^2W^2 + 2BW^3 + W^4.$$

Summing and dividing by 4 gives
$$B^4 + 2B^3W + 3B^2W^2 + 2BW^3 + W^4.$$

This is correct. The term B^4 corresponds to the all black colouring (which is in an equivalence class of its own); the term $2B^3W$ corresponds to the two equivalence classes with three black and one white side (one having a short white and one a long white side); the term $3B^2W^2$ corresponds to the three equivalence classes: long and short black, long and short white; two long blacks and two short whites; two short blacks and two long whites. The $2BW^3$ and W^4 terms similarly correspond correctly to two and one equivalence classes of colourings.

Solution 5.7
$$B^3 + G^3 + W^3 + B^2G + B^2W + G^2B + G^2W + W^2B + W^2G + BGW.$$

Solution 5.8

From Example 4.2, the cycle symbol for s is now
$$cs(s) = x_1 x_2.$$

Thus the colouring inventory of Fix(s) is
$$(R + Y + G + B)(R^2 + Y^2 + G^2 + B^2).$$

Solution 5.9

From Solution 4.9,
$$P_G(x_1, x_2) = \tfrac{1}{4}\left(x_1^4 + 2x_1^2 x_2 + x_2^2\right),$$

and so the pattern inventory is
$$\tfrac{1}{4}\left((B+G+W)^4 + 2(B+G+W)^2(B^2+G^2+W^2) + (B^2+G^2+W^2)^2\right).$$

Solution 5.10

From Solution 4.10,
$$P_G(x_1, x_2) = \tfrac{1}{4}\left(x_1^4 + 3x_2^2\right),$$

and so the pattern inventory is
$$\tfrac{1}{4}\left((B+G+W)^4 + 3(B^2+G^2+W^2)^2\right).$$

Solution 5.11

We must calculate $cs(g)$, for each g in D_3. Clearly, $cs(e) = x_1^6$, while each non-zero rotation creates a pair of 3-cycles, so that

$$cs(r) = cs(r^2) = x_3^2.$$

We have to be a little careful in analysing the reflection symmetries. In fact, the reflection axes are as shown below.

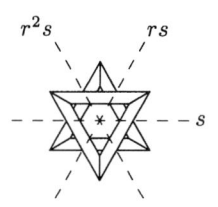

The lines through the vertices of the triangles and the centre are *not* reflection axes.

Thus, each reflection creates three 2-cycles, so that

$$cs(s) = cs(rs) = cs(r^2s) = x_2^3.$$

Adding the cycle symbols and dividing by 6 gives the cycle index:

$$P_G(x_1, x_2, x_3) = \tfrac{1}{6}\left(x_1^6 + 2x_3^2 + 3x_2^3\right).$$

Thus the pattern inventory is

$$\tfrac{1}{6}\left((S+G)^6 + 2(S^3+G^3)^2 + 3(S^2+G^2)^3\right)$$
$$= \tfrac{1}{6}\big(S^6 + 6S^5G + 15S^4G^2 + 20S^3G^3 + 15S^2G^4 + 6SG^5 + G^6$$
$$+ 2S^6 + 4S^3G^3 + 2G^6 + 3S^6 + 9S^4G^2 + 9S^2G^4 + 3G^6\big)$$
$$= S^6 + S^5G + 4S^4G^2 + 4S^3G^3 + 4S^2G^4 + SG^5 + S^6.$$

The total number of equivalence classes is $1+1+4+4+4+1+1 = 16$, of which 10 have at most three gold edges.

OBJECTIVES

After you have studied this unit, you should be able to:

(a) state the difference between individual colourings and equivalence classes of colourings of a configuration acted on by a symmetry group;

(b) recognize these equivalence classes as orbits under the action of the symmetry group on the colourings;

(c) distinguish between the group actions on the configuration positions and on the colourings;

(d) draw up the group action table for each of these actions;

(e) use the method of counting orbits based on summing the sizes of the stabilizers and dividing by the order of the group;

(f) appreciate the fact that this method stems from the Orbit–stabilizer Theorem;

(g) state the Counting Lemma, use it to count orbits under a group action, and use the m-colour version known as the Orbit Counting Theorem;

(h) in particular, use it in conjunction with the cycle index of the symmetry group, to count equivalence classes of colourings using the Cycle Index Theorem;

(i) state Pólya's Theorem and use it to obtain the pattern inventory of a configuration with a symmetry group action and a given set of colours.

INDEX

colouring inventory 28
configuration 10
counting lemma 19
cycle index 25
Cycle Index Theorem 26
cycle of length k 23
cycle symbol 23

equivalence classes 6
fixed set 18
group action 10
group action table 10
inert pairs 17
k-cycle 23
orbit 12

orbit–stabilizer Theorem 20
Orbit Counting Theorem 23
orbit number 16
Pólya's Theorem 31
pattern inventory 29
stabilizer 15